Notes on CATEGORIES AND GROUPOIDS

PHILIP. J. HIGGINS

King's College, London

VAN NOSTRAND REINHOLD COMPANY

VNR

LONDON

NEW YORK CINCINNATI TORONTO MELBOURNE

VAN NOSTRAND REINHOLD COMPANY
Windsor House, 46 Victoria Street, London, S.W.1

INTERNATIONAL OFFICES
New York Cincinnati Toronto Melbourne

Library of Congress Catalog Card No. 79–141979

SBN 442 03406 7

First published 1971

Printed in Great Britain by
Butler & Tanner Ltd
Frome and London

Preface

These notes are based on lectures given at the University of Michigan in 1967. Their aim is to present a self-contained account of the elementary theory of groupoids and some of its uses in group theory and topology. Category theory appears as a secondary topic whenever it is relevant to the main issue, and its treatment is by no means systematic. However, the book may serve as an introduction to categorical algebra with the emphasis always on specific applications.

One of my hopes in preparing the text was to convince students of group theory that it is often profitable to cross the boundary between groups and groupoids. The main advantage of the transition is that the category of groupoids provides a good model for certain aspects of homotopy theory. In it there are algebraic analogues of such notions as path, homotopy, deformation, covering and fibration. Most of these become vacuous when restricted to the category of groups, although they are clearly relevant to group-theoretical problems. This extra freedom is exploited in Chapters 14 and 15 to prove various sub-group theorems in the context of groupoids. There is another side to the coin: in applications of group theory to other topics it is often the case that the natural object of study is a groupoid rather than a group, and the algebra of groupoids may provide a more convenient tool for handling concrete problems. This point is illustrated in Chapter 17, where groupoids are used to compute fundamental groups. I hope that other applications will occur to the reader and that he will find the basic machinery he needs for them in these pages.

London, 1970. P. J. HIGGINS

CONTENTS

CHAPTER 1

Some basic categories

A *category* $\mathcal{K}$ consists of (i) a collection of *objects* $A, B, C, \ldots$, (ii) a family of disjoint sets $\mathcal{K}(A, B)$, one for each pair (A, B) of objects, (iii) a distinguished element ϵ_A of $\mathcal{K}(A, A)$ for each A, and (iv) a law of composition: if $\theta \in \mathcal{K}(A, B)$ and $\phi \in \mathcal{K}(B, C)$ then $\theta\phi \in \mathcal{K}(A, C)$; otherwise $\theta\phi$ is not defined. The members of $\mathcal{K}(A, B)$ are called $\mathcal{K}$-*morphisms* or $\mathcal{K}$-*maps* from A to B, and we write $\theta : A \longrightarrow B$ or $A \xrightarrow{\theta} B$ instead of $\theta \in \mathcal{K}(A, B)$ if it is clear from the context what category $\mathcal{K}$ is intended. The special element ϵ_A is called the identity morphism on A.

There are two axioms:

1. Associativity: if $A \xrightarrow{\theta} B \xrightarrow{\phi} C \xrightarrow{\psi} D$, then $(\theta\phi)\psi = \theta(\phi\psi)$.
2. Identity : if $A \xrightarrow{\theta} B$ then $\epsilon_A \theta = \theta = \theta \epsilon_B$.

If $\mathcal{K}$ and $\mathcal{L}$ are categories, then a *functor* $F : \mathcal{K} \to \mathcal{L}$ assigns to each object A of $\mathcal{K}$ an object $F(A)$ of $\mathcal{L}$, and to each $\mathcal{K}$-morphism $\alpha : A \to B$ an $\mathcal{L}$-morphism $F(\alpha) : F(A) \to F(B)$, in such a way that $F(\epsilon_A) = \epsilon_{F(A)}$ for each A, and $F(\alpha\beta) = F(\alpha)F(\beta)$ whenever $A \xrightarrow{\alpha} B \xrightarrow{\beta} C$.

The most familiar examples of categories are the *concrete* categories in which the objects are sets and the morphisms are certain mappings between them; composition is ordinary composition of mappings, and ϵ_A is the identity mapping on A. Usually the sets are provided with some extra structure (of a given type) and the morphisms are all the structure-preserving maps. Thus one has the category of rings and ring-homomorphisms, of topological groups and continuous homomorphisms, etc. Functors then arise naturally as "canonical constructions"; for example the following constructions all yield functors between the appropriate categories: (i) the integer group-ring $\mathbf{Z}(G)$ of a group G; (ii) the free group $F(X)$ on a set X; (iii) the fundamental group $\pi(T, *)$ of a topological space $(T, *)$ with base-point. (Note, however, that the centre of a group and the automorphism group of a group are not functorial constructions; always check what happens to the morphisms!)

The bulk of category theory aims at providing general theorems with applications in concrete categories (or in similar ones whose objects may be more complicated mathematical structures and whose morphisms may be, for example, families of mappings). Theorems of this type which appear in these notes will be concerned with such categorical notions as products, direct limits and adjoint functors, which describe in abstract terms some of the standard constructions occurring in many different branches of mathematics.

From a slightly different point of view one may also regard a category as an algebraic structure in its own right, on the same footing as a group or a semigroup. The "elements" of this algebra are the morphisms, and their composition is an associative partial binary operation; the objects act as labels for the morphisms in order to determine the domain of the operation. Functors are now to be thought of as structure-preserving maps, i.e. algebra homomorphisms. In the extreme case of a category with just one object, any two morphisms can be composed and the resulting algebra is a semigroup. Thus "category" is a direct generalisation of "semigroup-with-1". This algebraic view-point emphasises different aspects of category theory and leads to the consideration of such algebraic notions as congruences on a category, generators and relations for a category, free categories and word problems.

These two approaches to category theory will be developed side by side and will sometimes impinge on each other. For example, in the applications to group theory, categories will enter in both roles. A group G can be thought of as a category with one object whose morphisms are the elements of G. (This strain on the imagination is more fruitful than it appears at first sight; it is, in a sense, a reversion to the idea of a group of transformations). On the other hand, G is itself one of the objects of a category $\mathcal{G}_1$ whose morphisms are group homomorphisms. To avoid confusion in such situations we shall adopt a duplicate set of notations. When we are treating categories as abstract algebras we shall denote them by $A, B, C, \ldots$, and their morphisms by $a, b, x, y, \ldots$, the same notation as we use for groups and their elements. (The objects of such categories will only appear as suffixes attached to these symbols.) Algebra homomorphisms (functors) will be denoted by $\alpha, \beta, \gamma, \ldots$ in this notation, and they will be written as operators on the right. At other times, when we want to consider arbitrary categories (including "large" ones such as the category of all groups), we shall denote them by $\mathcal{A}, \mathcal{B}, \mathcal{C}, \ldots$, their objects by $A, B, C, \ldots$ and their morphisms by $\alpha, \beta, \gamma, \ldots$. We shall speak of *categories* and *Categories*, respectively, in these two contexts. The two notations occur simultaneously when the categories $A, B, C, \ldots$ are objects of the Category $\mathcal{C}$.

This informal distinction also serves as a set-theoretical warning. The reader can choose his own favourite set theory: if he likes to distinguish between sets and classes, he will interpret "category" (with a small c) to mean "small category", that is, one whose objects form a set; if he prefers to work with a hierarchy of universes he will interpret all categories as belonging to some fixed universe, but he may have to move to a larger universe when Categories appear.

We reserve special symbols for four Categories which will occur frequently.

1. *The Category* $\mathfrak{S}$ *of sets*. The objects of $\mathfrak{S}$ are sets and $\mathfrak{S}(A, B)$ is the set of all maps from A to B. Composition is the usual one, and ϵ_A is the identity map on A.

2. *The Category* $\mathfrak{D}$ *of directed graphs*. A *directed graph* A consists of (i) a set $V = V(A)$ of *vertices*, (ii) a set $E = E(A)$ of *edges*, and (iii) an incidence map $\delta : E \to V \times V$. If δ sends the edge x to the pair $x\delta = (x\delta_1, x\delta_2)$, we call $x\delta_1$ the *source* of x, and $x\delta_2$ the *target* of x. All graphs in these notes will be directed, and we call them simply "graphs".

A *graph-map* $\theta : A \to B$ is a pair of maps $V(\theta) : V(A) \to V(B)$ and $E(\theta) : E(A) \to E(B)$ which preserves incidences, i.e. such that $(x\,E(\theta))\delta_i = (x\delta_i)\,V(\theta)$ $(i = 1,2)$ for all edges x of A. The category $\mathfrak{D}$ has as its objects all (directed) graphs, and $\mathfrak{D}(A, B)$ is the set of all graph-maps from A to B Composition in $\mathfrak{D}$ is the obvious one obtained by composing the vertex maps and the edge maps separately. ϵ_A consists of the identity maps on $V(A)$ and $E(A)$.

We shall draw the usual pictures of graphs in which vertices are represented by points in the plane and edges by arrows from source to target. Since the vertices serve mainly as pegs to hang the edges on, we also adopt a notation in which they appear as labels. Writing I for $V(A)$, we let A_{ij} denote the set of edges from i to j (that is, $x \in A_{ij} \Longleftrightarrow x\delta = (i, j)$). Thus $E(A)$ is the disjoint union of sets A_{ij} $(i,j \in I)$, and we write $x : i \to j$ or $i \xrightarrow{x} j$ for $x \in A_{ij}$.

3. *The Category $\mathcal{C}$ of categories.* Any category has the structure of a directed graph with the objects as vertices and the morphisms as edges. In addition, the set of edges has a partial multiplication whose domain is determined by the graph structure. We may therefore make the following redefinition. A category A is a graph (with vertex set I and edge set $\bigcup A_{ij}$ as above), together with multiplications $A_{ij} \times A_{jk} \to A_{ik}$ for all $i, j, k \in I$, and distinguished edges $e_i \in A_{ii}$ $(i \in I)$. The product xy of edges is defined if and only if the source of y is the target of x. Multiplication is associative: ($(xy)z = x(yz)$ whenever one, hence the other, product is defined), and the e_i are identity elements ($e_i\, x = x = x\, e_j$ for $x \in A_{ij}$). In particular, each A_{ii} is a semigroup-with-1, the *vertex semigroup* at i.

A *category-map* is a graph map $\theta : A \to B$ between categories which preserves products and identity elements; in other words it is a functor from A to B. The Category $\mathcal{C}$ has as its objects all categories, and $\mathcal{C}(A, B)$ is the set of all category-maps from A to B. Composition in $\mathcal{C}$ is the same as in $\mathcal{D}$.

4. *The Category $\mathcal{G}$ of groupoids.* A *groupoid* A is a category in which every edge (morphism) is invertible; that is, every $x \in A_{ij}$ has an inverse $x^{-1} \in A_{ji}$ with $x x^{-1} = e_i$ and $x^{-1}x = e_j$. In particular, each A_{ii} is a group, the *vertex group* at i. If A and B are groupoids, a *groupoid-map* $\theta : A \to B$ is just a category-map from A to B. Such a category-map automatically preserves inverses; indeed in the presence of inverses one need only assume that θ preserves products. The Category $\mathcal{G}$ has as its objects all groupoids, and $\mathcal{G}(A, B)$ is the set of all groupoid-maps (functors) from A to B.

We have given these four Categories in order of increasing complexity of their objects, and we obtain "forgetful" functors

$\mathcal{G} \to \mathcal{C} \to \mathcal{D}$ by successively thinking of a groupoid as a category, and then merely as a graph. We also have two forgetful functors $V, E : \mathcal{D} \to \mathcal{S}$ which assign to each graph-map θ its *vertex map* $V(\theta)$ and its *edge map* $E(\theta)$. By abuse of notation we shall also use V, E to denote the obvious composite forgetful functors from $\mathcal{C}$ to $\mathcal{S}$ and from $\mathcal{G}$ to $\mathcal{S}$.

Useful variants of these Categories are the Categories $\mathcal{D}_I$, $\mathcal{C}_I$, $\mathcal{G}_I$ whose objects are all graphs (resp. categories, groupoids) with fixed vertex set I and whose morphisms are all those graph-maps (resp. category-maps, groupoid-maps) which leave all vertices fixed. We refer to I-categories, I-maps, etc. In particular, if I has just one member we obtain $\mathcal{C}_1$, the Category of semigroups-with-1 and $\mathcal{G}_1$, the Category of groups. $\mathcal{D}_1$ is indistinguishable from $\mathcal{S}$.

In all the above Categories one has obvious notions of sub-object, injection and surjection†. A morphism $\theta : A \to B$ in $\mathcal{D}$, $\mathcal{C}$ or $\mathcal{G}$ is an *injection* (resp. *surjection*) if both $V(\theta)$ and $E(\theta)$ are injections (resp. surjections) in $\mathcal{S}$. Clearly θ is an *isomorphism* (i.e. is invertible in the appropriate Category) if and only if it is both an injection and a surjection. The same applies to $\mathcal{D}_I$, $\mathcal{C}_I$, $\mathcal{G}_I$. If A and B are two objects of one of the Categories $\mathcal{D}$, $\mathcal{C}$, $\mathcal{G}$, $\mathcal{D}_I$, $\mathcal{C}_I$ or $\mathcal{G}_I$, we say that A is a *sub-object* (sub-graph, I-subcategory, etc.) of B if $A \subset B$ (i.e. $V(A) \subset V(B)$ and $E(A) \subset E(B)$) and the two inclusion maps form a morphism of that Category. We shall also speak of subgraphs of a groupoid, subgroupoids of a category, etc. in the obvious sense (apply a forgetful functor to one of the two

---

†*Injection* and *surjection* must not be confused with *monomorphism* and *epimorphism* which are defined in an arbitrary Category $\mathcal{A}$ as follows. $\alpha : A \to B$ is a monomorphism in $\mathcal{A}$ if, for all $\beta, \gamma : C \to A$ in $\mathcal{A}$, $\beta\alpha = \gamma\alpha$ implies $\beta = \gamma$. It is an epimorphism in $\mathcal{A}$ if, for all $\beta, \gamma : B \to C$ in $\mathcal{A}$, $\alpha\beta = \alpha\gamma$ implies $\beta = \gamma$. (See exercise 5 below).

objects). Thus a subgroupoid A of a category B is the same thing as a subcategory whose edges are all invertible (in A). A is a *full subgraph* of the graph B if it is obtained from B by selecting a set of vertices of B and including all edges of B whose source and target are both in this set (i.e. if $A_{ij} = B_{ij}$ for all $i,j \in V(A)$). A full subgraph of a category (groupoid) is always a subcategory (subgroupoid). *Note:* In a category or groupoid A, the vertex set $V(A)$ is essentially determined by the edge set $E(A)$ and the multiplication. (For the identity elements e are characterised by the property "$ex = x$, $ye = y$ whenever the products are defined", and there is just one at each vertex). It is therefore reasonable to think of A as a set of edges with multiplication and to write A for $E(A)$, $A = \bigcup A_{ij}$, $x \in A$ (where x is an edge) etc. Similarly if $\theta : A \to B$ is a category-map we may think of θ as a map of edge sets only and write θ for $E(\theta)$. With graphs, more care is required since there may be isolated vertices, and such notations would be inadmissible.

If $\theta : A \to B$ is a graph-map, we denote by $A\theta$ the image of A i.e. the subgraph of B formed by the images of the vertices and edges of A.

WARNING: if $\theta : A \to B$ is a category-map (or groupoid-map), the image $A\theta$ is not usually a subcategory (or subgroupoid) of B. (See exercise 1 below). We shall therefore usually be more interested in the subcategory or subgroupoid of B generated by $A\theta$, which is defined as follows. If $A^\lambda (\lambda \in \Lambda)$ are subgraphs of a graph B, their *intersection* $A = \bigcap_\lambda A^\lambda$ is the subgraph with $V(A) = \bigcap_\lambda V(A^\lambda)$ and $E(A) = \bigcap_\lambda E(A^\lambda)$. If B is a category, and each A^λ is a subcategory then their intersection is also a subcategory. Hence, for any subgraph X of a category B we may define cat $\{X\}$, *the subcategory generated by* X, to be the intersection of all subcategories containing X. Similarly if $X \subset B$ and B is a groupoid, gpd $\{X\}$ denotes the smallest subgroupoid of B containing X. Clearly $X \subset \text{cat}\,\{X\} \subset \text{gpd}\,\{X\}$ in this case.

PROPOSITION 1. *If $\theta : A \to B$ is a category-map (groupoid-map) and if its vertex map $V(\theta)$ is an injection, then the image $A\theta$ is a subcategory (subgroupoid) of B. In particular, this is true for all morphisms of $\mathcal{C}_I$ and $\mathcal{G}_I$.*

Proof. Let $b = a\theta$, $b' = a'\theta$ be edges of $A\theta$. If bb' is defined in $A\theta$, then $b\delta_2 = b'\delta_1$, so $a\delta_2 = a'\delta_1$ because of the condition on the vertex map. Hence $bb' = (aa')\theta \in A\theta$ and $A\theta$ is closed under the product in B. Also $A\theta$ contains the identity element at each of its vertices and, if A is a groupoid, contains inverses of all its edges. ■

PROPOSITION 2. *If the category (groupoid) A is generated by the subgraph X, then any category-map (groupoid-map) θ is uniquely determined by its restriction to X.*

Proof. The set of vertices and edges of A on which two category-maps agree is a subcategory of A. If two such maps agree on X they must therefore agree on the whole of A. The same argument works for groupoids. ■

To illustrate some of the ideas introduced above we take a simple but important example. Let I be any set and consider the graph $\Delta(I)$ whose vertex set is I and whose edge set is $I \times I$ (with δ the identity map on $I \times I$). $\Delta(I)$ has exactly one edge (i,j) from i to j for any $i,j \in I$, so there is a unique way of defining a category structure on $\Delta(I)$, namely by the rule $(i,j)(j,k) = (i,k)$. The edges (i,i) are the identity elements, and (j,i) is inverse to (i,j). Thus $\Delta(I)$ is a groupoid which we call a *simplicial groupoid,* and we shall always think of $\Delta(I)$ as carrying this structure. For any map

$\sigma : I \to J$, there is a unique groupoid map $\Delta(I) \to \Delta(J)$ with σ as its vertex map, and we see that Δ is a functor from $\mathcal{S}$ to $\mathcal{G}$. A subgraph of $\Delta(I)$ is essentially a binary relation on a subset of I; a subcategory of $\Delta(I)$ is a pre-ordering on a subset of I; a subgroupoid of $\Delta(I)$ is an equivalence relation on a subset of I. A *full* subgroupoid of $\Delta(I)$ is a groupoid $\Delta(J)$; an I-subgroupoid of $\Delta(I)$ is an equivalence relation on I.

We observe that if A is any I-graph, there is a canonical graph-map $\delta^* : A \to \Delta(I)$ given by the identity map on vertices and the incidence map $\delta : E(A) \to I \times I$ on edges. If A is a category (groupoid) then δ^* is a category-map (groupoid-map) and $A\delta^*$ is a subcategory (subgroupoid) of $\Delta(I)$ by Proposition 1. We call a graph A *unicursal* if there is at most one edge from i to j for each pair of vertices i, j. If A is a unicursal graph (category, groupoids), then δ^* is an injection, and A is isomorphic with a subgraph (subcategory, subgroupoid) of $\Delta(I)$. In particular, any groupoid (or category) with exactly one edge from i to j for each pair of vertices i, j is isomorphic with $\Delta(I)$, and we refer to all such groupoids as simplicial groupoids.

Exercises

1. Construct a groupoid map $\theta : A \to B$ such that $A\theta$ is not a subcategory of B.

2. Let X be a subgraph of the category A. Show that $a \in \text{cat}\,\{X\}$ (a an edge of A) if and only if either

(i) $a = e_i$ for some vertex i of X

or (ii) $a = x_1\, x_2 \ldots x_n$ for suitable edges $x_1, x_2, \ldots, x_n$ of X.

State and prove the corresponding result for groupoids.

3. Show that if $\theta : A \to B$ is a groupoid map, then the subcategory of B generated by $A\theta$ is a subgroupoid.

4. Let G be a groupoid. Show that the lattice of subgroupoids of G is a sublattice of the lattice of subcategories of G. (Note that these two lattices are not in general sublattices of the lattice of subgraphs of G, so there is something to prove).

5. Show that in the Categories $\mathcal{C}$ and $\mathcal{G}$, monomorphisms are the same as injections. Show also that in $\mathcal{C}$ or $\mathcal{G}$, if $\theta : A \to B$ is a morphism such that $A\theta$ generates B (as category or groupoid, respectively) then θ is an epimorphism. In $\mathcal{G}$ the converse is true, but is harder to prove. In $\mathcal{C}$ the converse is false; find a counter-example.

CHAPTER 2

Natural equivalence and adjoint functors

Let $\mathfrak{A}$, $\mathfrak{B}$ be Categories, and let F, G be functors from $\mathfrak{A}$ to $\mathfrak{B}$. A *natural transformation* τ from F to G (write $\tau : F \to G$) is a family of $\mathfrak{B}$-morphisms $\tau(A): F(A)$, one for each object A of $\mathfrak{A}$, with the property that, for every $\mathfrak{A}$-morphism $\alpha : A_1 \to A_2$, the following diagram commutes:

$$\begin{array}{ccc} F(A_1) & \xrightarrow{\tau(A_1)} & G(A_2) \\ {\scriptstyle F(\alpha)}\downarrow & & \downarrow{\scriptstyle G(\alpha)} \\ F(A_2) & \xrightarrow{\tau(A_2)} & G(A_2) \end{array}$$

Note. A diagram of objects and morphisms of a Category $\mathfrak{B}$ is called commutative if, whenever two morphisms are obtained by composing morphisms in the diagram along different routes between the same pair of endpoints, the two morphisms are equal. If the diagram is a subgraph X of $\mathfrak{B}$ (that is, if no object or morphism occurs more than once in the diagram), this condition says that the subcategory of $\mathfrak{B}$ generated by X is unicursal (see p.9, Exercise 2). In general, a *diagram* in $\mathfrak{B}$ is defined to be a graph-map $\theta : X \to \mathfrak{B}$ for some (directed) graph X, and such a diagram is called *commutative* if there exists a unicursal category C, a graph-map $\theta_1 : X \to C$, and a category-map $\theta_2 : C \to \mathfrak{B}$ with $\theta_1\, \theta_2 = \theta$ (See p.21, Exercise 1, and p.30, Exercise 1).

If $H : \mathfrak{A} \to \mathfrak{B}$ is another functor and $\sigma : G \to H$ a natural transformation, one obtains a composite natural transformation $\tau\sigma : F \to H$ by setting $(\tau\sigma)(A) = \tau(A)\sigma(A) : F(A) \to H(A)$. With the obvious definition of the identity transformation $F \to F$ one therefore obtains a Category $\mathfrak{B}^{\mathfrak{A}}$ whose objects are all the functors from $\mathfrak{A}$ to $\mathfrak{B}$ and whose morphisms are all † the natural transformations between them. A *natural equivalence* $\tau : F \to G$ is an invertible morphism of $\mathfrak{B}^{\mathfrak{A}}$, and it is easy to see that this is the same thing as a natural transformation τ such that $\tau(A) : F(A) \to G(A)$ is an isomorphism (in $\mathfrak{B}$) for all A. F and G are called *naturally equivalent functors* if there exists such a τ, and we write $F \simeq G$.

Examples. 1. The classical example, and one of the origins of category theory, concerns the second dual $F(A)$ of a finite dimensional vector space A over a field K. If $\mathfrak{V}$ is the category of such spaces and their linear maps, then F defines a functor from $\mathfrak{V}$ to $\mathfrak{V}$. The standard isomorphism $F(A) \to A$ for each A gives a natural equivalence $F \to 1$ (where 1 is the identity functor on $\mathfrak{V}$) and it is the existence of this natural equivalence which enables one to identify A with $F(A)$ for most purposes.

2. For any set X, let $F(X)$ denote the free group on X and $G(X)$ the free Abelian group on X. Then F and G define functors functors from $\mathfrak{S}$ to $\mathfrak{G}_1$, and the canonical homomorphisms $F(X) \to G(X)$ form a natural transformation $F \to G$.

3. For any directed graph A, let $E(A)$ be its set of edges and let $F(A) = V(A) \times V(A)$, where $V(A)$ is its set of vertices. Then E and F define functors from $\mathfrak{D}$ to $\mathfrak{S}$ and the incidence maps $\delta(A) : E(A) \to F(A)$ form a natural transformation $\delta : E \to F$.

† Here one may have to pass to a higher universe or restrict $\mathfrak{A}$ to be a small category. In the applications $\mathfrak{A}$ will always be suitably small.

Two Categories $\mathfrak{A}$ and $\mathfrak{B}$ are *isomorphic* ($\mathfrak{A} \cong \mathfrak{B}$) if there exist functors $F : \mathfrak{A} \to \mathfrak{B}$ and $G : \mathfrak{B} \to \mathfrak{A}$ such that $F \circ G = 1_{\mathfrak{B}}$ and $G \circ F = 1_{\mathfrak{A}}$. (Since we are writing functors on the left $F \circ G$ denotes "first G, then F". The symbol $\circ$ signals departure from our usual practice of composing maps the other way round). F and G are then *isomorphisms* of Categories. If, instead, one has only the weaker conditions $F \circ G \simeq 1$ and $G \circ F \simeq 1$ (natural equivalence of functors), one says that F and G are *equivalences* of Categories, and $\mathfrak{A}$ and $\mathfrak{B}$ are *equivalent* ($\mathfrak{A} \simeq \mathfrak{B}$). In most applications of Category theory one does not need to distinguish between equivalent Categories since the properties of greatest interest are preserved under equivalence. By the same token, naturally equivalent functors can be identified for most purposes. However, in the algebraic theory of categories there are interesting properties not preserved under equivalence. In fact, our main theme is the proposition that groupoids are often more useful than groups even though every groupoid is equivalent to a family of groups. For groupoids, the distinction between isomorphism and equivalence is, as we shall see, closely analogous to the distinction between homeomorphism and homotopy equivalence of topological spaces.

The following are some useful examples of isomorphisms of Categories.

1. If $\mathfrak{A}$ and $\mathfrak{B}$ are Categories, we define the *product* Category $\mathfrak{A} \times \mathfrak{B}$ in the obvious way. The objects are all pairs (A, B), where A is an object of $\mathfrak{A}$, and B is an object of $\mathfrak{B}$. The morphisms from (A_1, B_1) to (A_2, B_2) are all pairs (α, β), where $\alpha \in \mathfrak{A}\,(A_1, A_2)$ and $\beta \in \mathfrak{B}\,(B_1, B_2)$. Composition is defined by $(\alpha, \beta)(\alpha', \beta') = (\alpha\alpha', \beta\beta')$, and $\epsilon_{(A,B)} = (\epsilon_A, \epsilon_B)$. Clearly $\mathfrak{A} \times \mathfrak{B} \cong \mathfrak{B} \times \mathfrak{A}$, and $(\mathfrak{A}_1 \times \mathfrak{A}_2) \times \mathfrak{A}_3 = \mathfrak{A}_1 \times (\mathfrak{A}_2 \times \mathfrak{A}_3)$. Further, the product $\mathfrak{A}_1 \times \mathfrak{A}_2$ has

the usual universal property : any pair of functors (F_1, F_2), $F_i : \mathcal{B} \to \mathcal{A}_i$ gives rise to a unique functor $F : \mathcal{B} \to \mathcal{A}_1 \times \mathcal{A}_2$ which, when composed with the projection functors $\mathcal{A}_1 \times \mathcal{A}_2 \to \mathcal{A}_1$ and $\mathcal{A}_1 \times \mathcal{A}_2 \to \mathcal{A}_2$ gives back the functors F_1 and F_2. This fact may be stated as an isomorphism of Categories $(\mathcal{A}_1 \times \mathcal{A}_2)^{\mathcal{B}} \cong \mathcal{A}_1^{\mathcal{B}} \times \mathcal{A}_2^{\mathcal{B}}$, which requires some verification as regards the *morphisms* of the Categories (which are natural transformations). We leave this to the reader, who should also convince himself that $\mathcal{A}^{\mathcal{B}_1 \times \mathcal{B}_2} \cong ((\mathcal{A})^{\mathcal{B}_1})^{\mathcal{B}_2} \cong ((\mathcal{A})^{\mathcal{B}_2})^{\mathcal{B}_1}$. All these isomorphisms are canonical (in a sense which would have to be described by natural transformations in a higher universe) and we shall use them to identify the Categories in question.

2. If $\mathcal{A}$ is any Category, we define the *dual* or *opposite* Category $\mathcal{A}^{\mathrm{op}}$ as follows. The objects A^{op} of $\mathcal{A}^{\mathrm{op}}$ are in one-one correspondence with the objects A of $\mathcal{A}$. The morphisms $\theta^{\mathrm{op}} : A^{\mathrm{op}} \to B^{\mathrm{op}}$ are in one-one correspondence with the morphisms $\theta : B \to A$, and the composition is defined by $\theta^{\mathrm{op}} \phi^{\mathrm{op}} = (\phi\theta)^{\mathrm{op}}$. (One can, for example, take the objects and morphisms of $\mathcal{A}$ itself with the oppositely oriented graph structure and the opposite multiplication, but this is liable to lead to confusion). There are canonical isomorphisms $(\mathcal{A}^{\mathrm{op}})^{\mathrm{op}} \cong \mathcal{A}$, $(\mathcal{A}^{\mathrm{op}} \times \mathcal{B}^{\mathrm{op}}) \cong (\mathcal{A} \times \mathcal{B})^{\mathrm{op}}$ and $(\mathcal{A}^{\mathrm{op}})^{\mathcal{B}^{\mathrm{op}}} \cong (\mathcal{A}^{\mathcal{B}})^{\mathrm{op}}$. The reader should check the last of these to see that the natural transformations go in the right direction.

The objects of $\mathcal{A}^{\mathcal{B}^{\mathrm{op}}}$ (functors from $\mathcal{B}^{\mathrm{op}}$ to $\mathcal{A}$) are called *contravariant functors* from $\mathcal{B}$ to $\mathcal{A}$. They can be thought of as maps from $\mathcal{B}$ to $\mathcal{A}$ which reverse the graph structure and the multiplication, i.e. anti-homomorphisms. The usual notation encourages this attitude: we write $F : \mathcal{B} \to \mathcal{A}$, and $F(\beta) : F(B_1) \to F(B_2)$, where

$\beta : B_2 \to B_1$ in $\mathcal{B}$. Natural transformations between such functors are defined as morphisms in $\mathcal{A}^{\mathcal{B}^{op}}$ (not morphisms in $(\mathcal{A}^{op})^{\mathcal{B}}$, which go the other way). The objects of $\mathcal{A}^{\mathcal{B}}$ should now be renamed *covariant functors*, but we shall not often use this term; functors are always covariant unless otherwise stated.

There are other extensions of this terminology. For example, a functor $F : \mathcal{A}^{op} \times \mathcal{B} \to \mathcal{E}$ is described as a functor of two variables, contravariant in the first and covariant in the second. If $\alpha \in \mathcal{A}(A_2, A_1)$ and $\beta \in \mathcal{B}(B_1, B_2)$ then $F(\alpha,\beta) : F(A_1, B_1) \to F(A_2, B_2)$ in $\mathcal{E}$. The standard example of this situation is the functor $H_{\mathcal{A}} : \mathcal{A}^{op} \times \mathcal{A} \to \mathcal{S}$ (for any Category $\mathcal{A}$) given by $H_{\mathcal{A}}(A,B) = \mathcal{A}(A,B)$. For $\alpha : A_2 \to A_1$ and $\beta : B_1 \to B_2$ in $\mathcal{A}$, the corresponding map $H_{\mathcal{A}}(\alpha,\beta) : \mathcal{A}(A_1,B_1) \to \mathcal{A}(A_2,B_2)$ sends σ to $\alpha\sigma\beta$.

Suppose now that $F : \mathcal{A} \to \mathcal{B}$ and $G : \mathcal{B} \to \mathcal{A}$ are (covariant) functors. Then, by composition with the functors $H_{\mathcal{A}}$ and $H_{\mathcal{B}}$ above, we obtain two functors $P, Q : \mathcal{A}^{op} \times \mathcal{B} \to \mathcal{S}$ which are given on objects by $P(A,B) = \mathcal{B}(F(A), B)$ and $Q(A,B) = \mathcal{A}(A, G(B))$. Their effect on morphisms $\alpha : A_2 \to A_1$ and $\beta : B_1 \to B_2$ is described as follows: $P(\alpha,\beta)$ is the map from $\mathcal{B}(F(A_1),B_1)$ to $\mathcal{B}(F(A_2),B_2)$ sending ρ to $F(\alpha)\rho\beta$; $Q(\alpha,\beta)$ is the map from $\mathcal{A}(A_1, G(B_1))$ to $\mathcal{A}(A_2, G(B_2))$ sending σ to $\alpha\sigma G(\beta)$. If the two functors P and Q are naturally equivalent, we say that (F,G) is an *adjoint pair* of functors. This relation is not symmetric: we say that F is *left adjoint* to G and that G is *right adjoint* to F. We also say that F *is* a left adjoint and *has* a right adjoint, etc. As usual in Category theory, the definition is much simpler than it seems at first sight. It says, first, that there is a one-one correspondence between morphisms $\sigma : A \to G(B)$ in $\mathcal{A}$ and morphisms $\sigma^* : F(A) \to B$ in $\mathcal{B}$ and, second, that the correspondences $\sigma \longleftrightarrow \sigma^*$ can be chosen simultaneously for all A, B so that they are natural with respect to the morphisms of $\mathcal{A}$ and $\mathcal{B}$, which means that whenever one of the diagrams

$$\begin{array}{ccc} A_1 & \xrightarrow{\sigma} & G(B_1) \\ {\scriptstyle \alpha}\uparrow & & \downarrow{\scriptstyle G(\beta)} \\ A_2 & \xrightarrow{\tau} & G(B_2) \end{array} \qquad \begin{array}{ccc} F(A_1) & \xrightarrow{\sigma^*} & B_1 \\ {\scriptstyle F(\alpha)}\uparrow & & \downarrow{\scriptstyle \beta} \\ F(A_2) & \xrightarrow{\tau^*} & B_2 \end{array}$$

commutes, so does the other.

One deduces immediately that if $\alpha_i : A \to A_i$, $\sigma_i : A_i \to G(B_i)$ and $\beta_i : B_i \to B$ $(i = 1,2)$, then
$\alpha_1 \sigma_1 G(\beta_1) = \alpha_2 \sigma_2 G(\beta_2) \Longleftrightarrow F(\alpha_1)\, \sigma_1^* \beta_1 = F(\alpha_2)\, \sigma_2^* \beta_2$. In particular, if one of the diagrams

$$\begin{array}{ccc} A_1 & \xrightarrow{\sigma_1} & G(B_1) \\ {\scriptstyle \alpha}\downarrow & & \downarrow{\scriptstyle G(\beta)} \\ A_2 & \xrightarrow[\sigma_2]{} & G(B_2) \end{array} \qquad \begin{array}{ccc} F(A_1) & \xrightarrow{\sigma_1^*} & B_1 \\ {\scriptstyle F(\alpha)}\downarrow & & \downarrow{\scriptstyle \beta} \\ F(A_2) & \xrightarrow[\sigma_2^*]{} & B_2 \end{array}$$

commutes, then so does the other.

Some familiar examples should make this concept clear.

1. Let $\mathcal{A} = \mathcal{S}$ and $\mathcal{B} = \mathcal{G}_1$, the Category of groups. Let $F : \mathcal{S} \to \mathcal{G}_1$ assign to each set A the free group $F(A)$ on A. Let $G : \mathcal{G}_1 \to \mathcal{S}$ assign to each group B its underlying set $G(B)$. Every map from A to the set $G(B)$ extends uniquely to a group homomorphism $F(A) \to B$, and this gives a one-one correspondence $\mathcal{S}(A, G(B)) \to \mathcal{G}_1(F(A), B)$, which is clearly natural (a composite map $A' \to A \to G(B)$ extends to the composite homomorphism

$F(A') \to F(A) \to B$ etc.). Thus the free group functor $\mathcal{S} \to \mathcal{G}_1$ is left adjoint to the forgetful functor $\mathcal{G}_1 \to \mathcal{S}$.

2. More typically, let $\mathcal{A} = \mathcal{G}_1$, and let $\mathcal{B} = \mathcal{R}$, the Category of rings-with-1. For any group A, let $F(A)$ be its group ring over the integers. For any ring B, let $G(B)$ be its group of units. Then we have functors $F : \mathcal{G}_1 \to \mathcal{R}$ and $G : \mathcal{R} \to \mathcal{G}_1$ with the obvious effect on morphisms. Now every group homomorphism from A to the units of B induces a ring homomorphism from the group ring of A to B. The converse is also true (we require, of course, that the morphisms in $\mathcal{R}$ preserve the 1) and we have a natural one-one correspondence between $\mathcal{G}_1(A, G(B))$ and $\mathcal{R}(F(A), B)$. Thus (F, G) is an adjoint pair.

3. Let $\mathcal{A} = \mathcal{B} = \mathcal{T}^*$, the Category of topological spaces with base-point. Let $S(A)$ denote the reduced suspension of A, and $\Omega(B)$ the loop space of B. Then (S, Ω) is an adjoint pair of functors from $\mathcal{T}^*$ to $\mathcal{T}^*$.

4. An example from Ch. 1. The forgetful functor $V : \mathcal{G} \to \mathcal{S}$ ($V(G)$ is the set of vertices of the groupoid G) has as right adjoint the functor $\Delta : \mathcal{S} \to \mathcal{G}$. ($\Delta(I)$ is the simplicial groupoid with vertex set I). V also has a left adjoint, namely the functor $T : \mathcal{S} \to \mathcal{G}$ which assigns to each set I the "trivial" groupoid $T(I)$ with vertex set I and no edges apart from the identity elements $e_i (i \in I)$. Thus (V, Δ) and (T, V) are adjoint pairs. There is also an adjoint pair (C, T), as we shall see in the next chapter, but there is no adjoint pair $(\Delta, ?)$.

We shall prove in a moment that if a functor F has a right (or left) adjoint, then that adjoint is uniquely determined by F up to natural equivalence. Thus, for example, the description "left adjoint of the forgetful functor from groups to sets" characterises the construction of free groups (up to isomorphism). The language of adjoints provides in this way a very convenient method of describing the universal properties of such constructions, and we shall make use of it throughout the following pages. We hasten to add that convenience is not the only reason for introducing adjoints, for they play a vital role in much of Category theory. Their most important property (preservation of limits) will be discussed in Ch.7, and will provide a unified treatment of several later topics.

The connection between adjoints and universal properties arises as follows. Suppose that $F : \mathcal{A} \to \mathcal{B}$ and $G : \mathcal{B} \to \mathcal{A}$ form an adjoint pair (F, G), and choose fixed isomorphisms $\mathcal{A}(A, G(B)) \cong \mathcal{B}(F(A), B)$ which together form a natural equivalence. Putting $B = F(A)$ we have, for each A, an isomorphism of sets $\mathcal{A}(A, G(F(A))) \cong \mathcal{B}(F(A), F(A))$. This enables us to pick out a special $\mathcal{A}$-morphism $\sigma_A : A \to G(F(A))$ which corresponds to the identity morphism $\epsilon_{F(A)}$ in $\mathcal{B}$.
Of the two corresponding diagrams

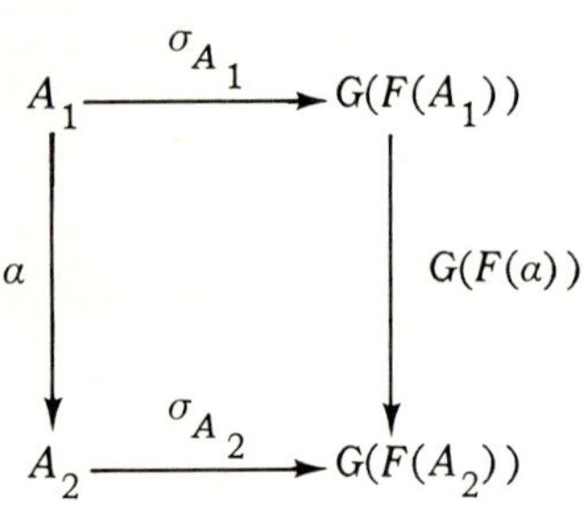

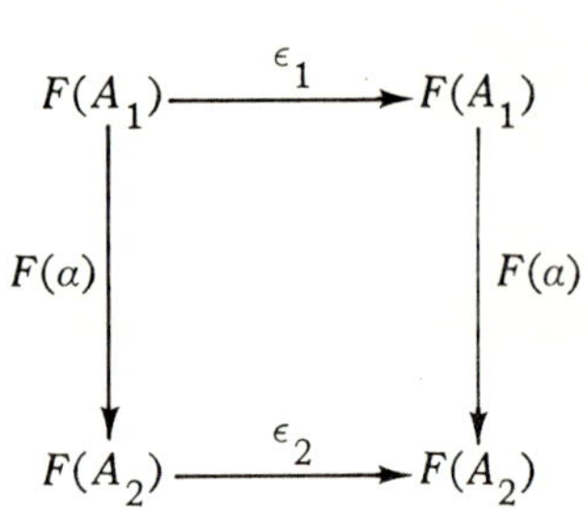

the second commutes; therefore so does the first, i.e. the σ_A form a natural transformation $\sigma : 1_{\mathcal{A}} \to G \circ F$. Suppose now that an $\mathcal{A}$-morphism $\alpha : A \to G(B)$ is given, and consider corresponding diagrams

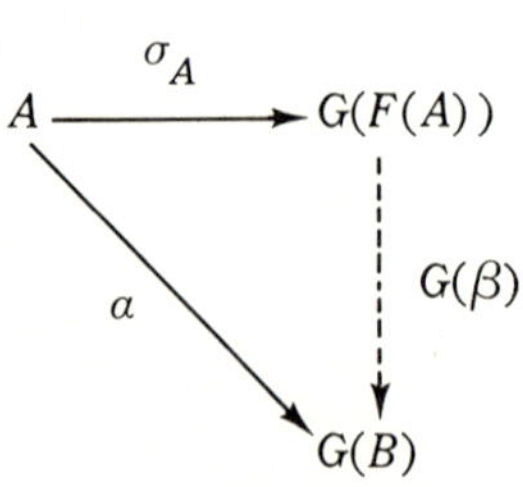

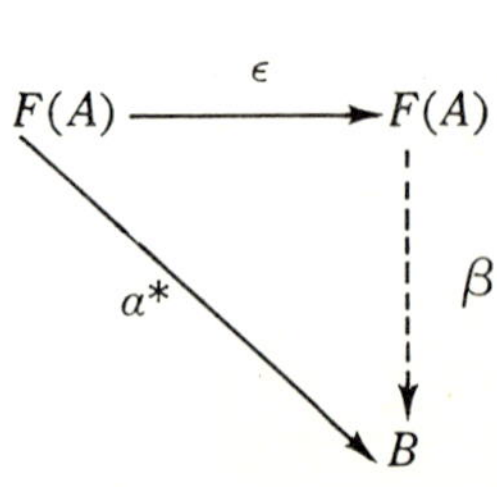

The second commutes if and only if $\beta = \alpha^*$; therefore there is one and only one morphism of the form $G(\beta)$ which makes the first commute, namely $G(\alpha^*)$. In this sense σ_A is universal for $\mathfrak{A}$-morphisms $\alpha : A \to G(B)$.

PROPOSITION 3. *Let $F : \mathfrak{A} \to \mathfrak{B}$ and $G : \mathfrak{B} \to \mathfrak{A}$ be functors. Then the following are equivalent:*

(i) (F, G) is an adjoint pair.

(ii) $\exists$ $\mathfrak{A}$-morphisms $\sigma_A : A \to G(F(A))$ such that

(a) $\sigma : 1_{\mathfrak{A}} \to G \circ F$ is a natural transformation, and

(b) for every $\mathfrak{A}$-morphism $\alpha : A \to G(B)$ there is a unique $\mathfrak{B}$-morphism $\beta : F(A) \to B$ such that $\sigma_A G(\beta) = \alpha$,

(iii) $\exists$ $\mathfrak{B}$-morphisms $\tau_B : F(G(B)) \to B$ such that

(a) $\tau : F \circ G \to 1_{\mathfrak{B}}$ is a natural transformation and

(b) for every $\mathfrak{B}$-morphism $\beta : F(A) \to B$ there is a unique $\mathfrak{A}$-morphism $\alpha : A \to G(B)$ such that $F(\alpha)\tau_B = \beta$.

Proof. (i) $\Longrightarrow$ (ii) has been proved above.

(ii) $\Longrightarrow$ (i). The existence and uniqueness of β gives a map $\alpha \mapsto \beta$ from $\mathfrak{A}(A, G(B))$ to $\mathfrak{B}(F(A), B)$. There is a map the other way given by $\beta \mapsto \sigma_A G(\beta)$, and condition (b) ensures that these are inverse maps. We write $\beta = \alpha^*$. The naturality of σ implies that for $\lambda : A_1 \to A$, $\lambda\sigma_A = \sigma_{A_1} G(F(\lambda))$. Hence, for $\alpha : A \to G(B)$ and $\mu : B \to B_1$, we have $\lambda\alpha G(\mu) = \lambda\sigma_A G(\alpha^*)G(\mu) = \sigma_{A_1} G(F(\lambda))G(\alpha^*)G(\mu) = \sigma_{A_1} G(F(\lambda)\alpha^*\mu)$. Thus $(\lambda\alpha G(\mu))^* = F(\lambda)\alpha^*\mu$, which is the condition for $\alpha \mapsto \alpha^*$ to be natural.

(i) $\Longleftrightarrow$ (iii) follows by duality. (Think of F and G as functors $\overline{F} : \mathfrak{A}^{op} \to \mathfrak{B}^{op}$ and $\overline{G} : \mathfrak{B}^{op} \to \mathfrak{A}^{op}$. Then (F, G) is an adjoint pair if and only if $(\overline{G}, \overline{F})$ is an adjoint pair). ∎

PROPOSITION 4. *(i) Let (F, G), (F', G') be adjoint pairs of functors* *with* $F, F' : \mathfrak{A} \to \mathfrak{B}$ *and* $G, G' : \mathfrak{B} \to \mathfrak{A}$. *Then* $F \simeq F'$ *if and only if* $G \simeq G'$. *(Thus F determines G to within natural equivalence, and vice versa).*

(ii) If (F_1, G_1) *and* (F_2, G_2) *are adjoint pairs with* $\mathfrak{A} \xrightarrow{F_1} \mathfrak{B} \xrightarrow{F_2} \mathfrak{C}$ *and* $\mathfrak{C} \xrightarrow{G_2} \mathfrak{B} \xrightarrow{G_1} \mathfrak{A}$, *then* $(F_2 \circ F_1, G_1 \circ G_2)$ *is an adjoint pair.*

Proof. (i) Suppose that $G \simeq G'$. Then (F', G) is an adjoint pair since the induced isomorphisms $\mathfrak{A}(A, G(B)) \simeq \mathfrak{A}(A, G'(B)) \simeq \mathfrak{B}(F'(A), B)$ form natural transformations. We may therefore assume that $G' = G$. Denote by σ_A and σ'_A the special $\mathfrak{A}$-morphisms $A \to G(F(A))$ and $A \to G(F'(A))$ as above. By the universal property of σ_A there is, for each A, a unique $\mathfrak{B}$-morphism $\beta_A : F(A) \to F'(A)$ such that $\sigma_A G(\beta_A) = \sigma'_A$. In the diagram

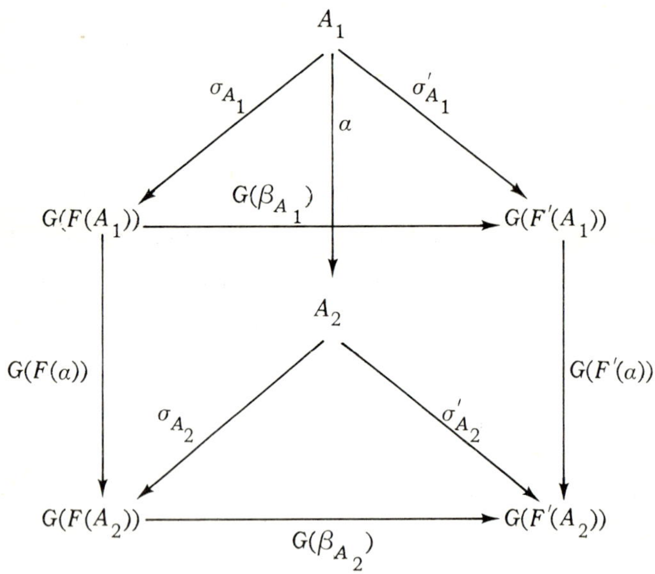

the left and right parallelograms commute (since σ, σ' are natural) and the two triangles commute (by definition of β_{A_1} and β_{A_2}). The map $\theta = \alpha\, \sigma'_{A_2} : A_1 \to G(F'(A_2))$ can therefore be written as either

$\sigma_{A_1} G(\beta_{A_1}) G(F'(\alpha))$ or $\sigma_{A_1} G(F(\alpha)) G(\beta_{A_2})$; in other words, the equation $\theta = \sigma_{A_1} G(\beta)$ has the two solutions $\beta = \beta_{A_1} F'(\alpha)$ and $\beta = F(\alpha)\beta_{A_2}$. It follows that $\beta_{A_1} F'(\alpha) = F(\alpha)\beta_{A_2}$, that is, the β_A form a natural transformation $\beta : F \to F'$. In fact β is the unique transformation from F to F' satisfying $\sigma_A G(\beta_A) = \sigma'_A$ for all A. Similarly, there is a unique natural transformation $\gamma : F' \to F$ satisfying $\sigma'_A G(\gamma_A) = \sigma_A$. The composite transformation $\beta\gamma : F \to F$ is then the unique natural transformation satisfying $\sigma_A G((\beta\gamma)_A) = \sigma_A$, namely the identity transformation on F, and $\gamma\beta$ is the identity transformation on F'. Thus $F \simeq F'$, as claimed. The converse follows by duality.

(ii) There are natural isomorphisms $\mathfrak{A}(A, G_1(B)) \simeq \mathfrak{B}(F_1(A), B)$ and $\mathfrak{B}(B', G_2(C)) \simeq \mathfrak{C}(F_2(B'), C)$. On restricting these to the situation $B = G_2(C)$ and $B' = F_1(A)$ they can be composed to give isomorphisms $\mathfrak{A}(A, G_1(G_2(C))) \simeq \mathfrak{C}(F_2(F_1(A)), C)$ which form a natural equivalence. ■

Exercises

1. Let X be a graph and $\theta : X \to \mathfrak{B}$ a diagram in the Category $\mathfrak{B}$. Let X^* denote the image of X under the canonical map $\delta^* : X \to \Delta(I)$, where $I = V(X)$. Writing δ^* also for the induced map $X \to X^*$, show that the diagram is commutative in the sense defined above if and only if $\theta = \delta^*\phi$ for some category-map $\phi : X^* \to \mathfrak{B}$.

2. Let $F : \mathfrak{A} \to \mathfrak{B}$ and $G : \mathfrak{B} \to \mathfrak{A}$ be functors. Show that if both (F, G) and (G, F) are adjoint pairs then $F \circ G \circ F \simeq F$ and $G \circ F \circ G \simeq G$.

3. Let $E_1 : \mathfrak{D} \to \mathfrak{S}$, $E_2 : \mathfrak{C} \to \mathfrak{S}$, $E_3 : \mathfrak{G} \to \mathfrak{S}$ denote the forgetful functors associating to each graph, category or groupoid its set of edges. Find left adjoints for each of E_1, E_2, E_3.

4. Show that if $F : \mathfrak{A} \to \mathfrak{B}$ has a right adjoint then F preserves epimorphisms (i.e. if α is an epimorphism in $\mathfrak{A}$ then $F(\alpha)$ is an epimorphism in $\mathfrak{B}$). Dually, any right adjoint functor preserves monomorphisms.

CHAPTER 3

Paths and components

The existence of free algebras and presentations of algebras by generators and relations depends on the notion of *words* or *well-formed formulae.* These are formulae containing variables and operators so constructed that they make sense, i.e. so that when the variables are replaced by elements of an appropriate algebra the word can be evaluated by calculation in the algebra. In the case of semigroups it is usual to omit the symbol for multiplication and also the brackets and define a *semigroup-word* to be just a string of variables, and the introduction of inverses of variables leads to the notion of group-words. We want to generalise these ideas to the case of categories and groupoids, and since multiplication is not always defined in these algebras we must somehow specify when two variables can be multiplied. To do this we take for our variables the set of edges of a fixed graph, and multiply two edges if and only if they abut. The resulting *category-words* are essentially the same as directed paths in the graph, and groupoid-words are undirected paths. In this context "path" and "word" are almost synonymous, and we shall see that the application of standard arguments to paths instead of words leads to the construction of free categories and free groupoids and to the solution of the word problem for these algebras. Our first task is to make these notions precise.

Let $[n]$ denote the graph

with $n+1$ vertices and n edges joining them in sequence ($n \geqslant 0$). If X is any (directed) graph and i, j are vertices of X, we define a *directed path of length n from i to j in X* to be a graph map $p : [n] \to X$ with $0 \mapsto i$ and $n \mapsto j$. In particular, there is one directed path of length 0 from i to i for each vertex i of X. Equivalently, one may think of a directed path as a sequence $p = (x_1, x_2, \dots, x_n)$ of edges of X which link up in sequence. (There is an ambiguity in this notation when $n = 0$ and one should write $(\ )_i$ or $\emptyset_i$ for the "empty" path at the vertex i). If $p = (x_1, x_2, \dots, x_n)$ and $q = (y_1, y_2, \dots, y_m)$ are directed paths from i to j and from j to k, respectively, then $pq = (x_1, x_2, \dots, x_n, y_1, \dots, y_m)$ is a directed path from i to k. Thus we have a multiplication of paths : $\vec{P}_{ij} \times \vec{P}_{jk} \to \vec{P}_{ik}$, where $\vec{P}_{ij}$ denotes the set of all directed paths (of all lengths) from i to j. This multiplication is associative, so we obtain a category $\vec{P}(X)$, the *category of directed paths in X*. Its vertex set is $V(X)$ and its identities are the paths of zero length.

If $\theta : X \to Y$ is a graph-map, and $p : [n] \to X$ is a directed path from i to j in X, then $p\theta : [n] \to Y$ is a directed path from $i\theta$ to $j\theta$ in Y. Thus θ induces a graph-map $\vec{P}(\theta) : \vec{P}(X) \to \vec{P}(Y)$, and it is easy to check that it is actually a category-map. This shows that we have constructed a functor $\vec{P} : \mathfrak{D} \to \mathfrak{C}$. We also obtain a canonical embedding $\sigma_X : X \to \vec{P}(X)$ by considering each edge of X as a path of length 1. σ_X is a graph-map, and we shall often identify X with its image in $\vec{P}(X)$. Finally, if A is a category, and $F : \mathfrak{C} \to \mathfrak{D}$ is the forgetful functor, consider the category $\vec{P}(F(A))$ of directed paths in $F(A)$. If $p = (a_1, a_2, \dots, a_n)$ is a path in $F(A)$ then the product

$a = a_1 a_2 \ldots a_n$ is defined in A, and we call it the *value* of p in A. (The value of the empty word at i is defined to be the identity e_i of A). We therefore obtain an *evaluation map* $\epsilon_A : \vec{P}(F(A)) \to A$ defined by $p\,\epsilon_A$ = (value of p in A), and it is clear from the definition of multiplication for paths that ϵ_A is a category-map.

PROPOSITION 5. *$\vec{P} : \mathfrak{D} \to \mathfrak{C}$ is left adjoint to the forgetful functor $F : \mathfrak{C} \to \mathfrak{D}$.*

Proof. The canonical maps $\sigma_X : X \to F(\vec{P}(X))$ form a natural transformation $\sigma : 1_{\mathfrak{D}} \to F \circ \vec{P}$. If $\alpha : X \to F(A)$ is a graph-map from X to a category A, then $\vec{P}(\alpha) : \vec{P}(X) \to \vec{P}(F(A))$ can be composed with the evaluation map ϵ_A to give a category-map $\beta = \vec{P}(\alpha)\epsilon_A : \vec{P}(X) \to A$. The diagram

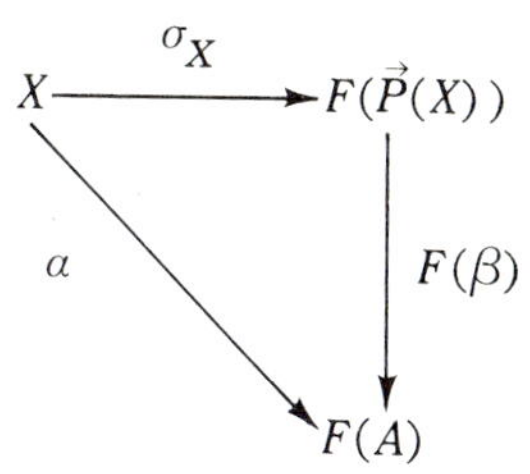

commutes, since the restriction of β to (the image of) X in $\vec{P}(X)$ is just α. Also, β is unique with this property (by Proposition 2) since X generates $\vec{P}(X)$. The result follows now from Proposition 3. ∎

The universal property of $\vec{P}(X)$ described above reduces to the universal property of free semigroups when X has just one vertex. We therefore define free categories as follows. Let X be a given graph and $\sigma : X \to C$ a graph-map into a category C. We say that C is the *free category* on X (relative to σ) if for every graph-map $\theta : X \to A$, where A is a category, there is a unique category-map

$\theta^* : C \to A$ such that $\theta = \sigma\theta^*$. This property determines C and σ up to isomorphism, and the proof of Proposition 5 shows that there is a free category on any graph X, namely $\vec{P}(X)$.

We now need the notion of (undirected) *path* in a graph X. Intuitively this means a map of the undirected graph •——•——• .... •——• into the undirected graph corresponding to X. This would be an adequate definition in some contexts, but here we need to be more careful since we want to use these paths in place of group-words and it is essential that to every edge in X from i to i there should correspond *two* paths of length 1 from i to i (one in each direction, so to speak). We arrange for this as follows:

An *involution of a graph* X is a map $E(X) \to E(X)$ which sends each $x \in X_{ij}$ to some $\bar{x} \in X_{ji}$ in such a way that $\bar{\bar{x}} = x$ for all $x \in E(X)$. If X and Y are graphs provided with fixed involutions $x \mapsto \bar{x}$, $y \mapsto \bar{y}$ then a *morphism of graphs-with-involution* is a graph-map $\alpha : X \to Y$ such that $\overline{x\alpha} = \bar{x}\alpha$ for all $x \in E(X)$. This defines the Category $\overline{\mathfrak{D}}$ of graphs-with-involution. Similarly one can define the Category $\overline{\mathfrak{C}}$ of categories-with-involution (an involution of a category A is a contravariant functor $A \to A$ of order 2). Now to each graph X one can associate a graph-with-involution $\overline{X}$ as follows. Let X^{op} be a graph anti-isomorphic to X and having no edges in common with X, and let $\bar{x}$ denote the element of X^{op}_{ji} corresponding to $x \in X_{ij}$. Define $\overline{X}$ to be the graph with the same vertices as X but with edges $\overline{X}_{ij} = X_{ij} \cup X^{\mathrm{op}}_{ij}$ for all i, j. The map $x \mapsto \bar{x}$, $\bar{x} \mapsto x$ is an involution of $\overline{X}$ and we define $\bar{\bar{x}} = x$ for $x \in E(X)$. This construction obviously gives a functor $J : \mathfrak{D} \to \overline{\mathfrak{D}}$ and the reader should check that it is left adjoint to the forgetful functor $\overline{\mathfrak{D}} \to \mathfrak{D}$.

We now define a *path in* X to be a directed path in $\overline{X}$ and denote by $P(X) = \vec{P}(\overline{X})$ the *category of paths in* X. There are now two paths x, $\bar{x}$ of length 1 for each edge of X, but still only one path of zero length at each vertex. If $p = (y_1, y_2, \ldots, y_n)$ is a path from i to j in X $(y_i \in E(\overline{X}))$, then $\overline{p} = (\overline{y}_n, \ldots, \overline{y}_2, \overline{y}_1)$ is a path from j to i, and clearly $p \mapsto \overline{p}$ is an involution of the category $P(X)$. Thus P is a functor from $\mathfrak{D}$ to $\overline{\mathcal{C}}$.

PROPOSITION 5′. *$P : \mathfrak{D} \to \overline{\mathcal{C}}$ is left adjoint to the forgetful functor $\overline{\mathcal{C}} \to \mathfrak{D}$.*

Proof. In the proof of Proposition 5, if X and A are provided with involutions, and all maps are required to preserve involutions, the same argument shows that the induced functor $\vec{P} : \overline{\mathfrak{D}} \to \overline{\mathcal{C}}$ is left adjoint to the forgetful functor $\overline{\mathcal{C}} \to \overline{\mathfrak{D}}$. Now $P = \vec{P} \circ J$, where $J : \mathfrak{D} \to \overline{\mathfrak{D}}$ is left adjoint to the forgetful functor $\overline{\mathfrak{D}} \to \mathfrak{D}$. Hence P is left adjoint to the composite forgetful functor $\overline{\mathcal{C}} \to \mathfrak{D}$ by Proposition 4 (ii). ■

DEFINITIONS. A graph X is *connected* if there is at least one path from i to j in X for each pair of vertices i, j. A category or groupoid is *connected* if its underlying graph is connected. A graph X is the *disjoint union* of subgraphs X^λ $(\lambda \in \Lambda)$ if $V(X)$ and $E(X)$ are, respectively, the disjoint union of their subsets $V(X^\lambda)$ and $E(X^\lambda)$; the X^λ are then full subgraphs of X.

A maximal connected subgraph of X is called a *connected component* of X, or simply a *component* of X. If one defines a relation $i \sim j$ on the vertices of X by the rule "$i \sim j$ if there is a path from i to j" then, because $P(X)$ is a category with involution, $\sim$ is an equivalence relation, and it is clear that the components of X are just the full subgraphs on the equivalence classes of vertices. We leave the reader to verify to following simple facts:

PROPOSITION 6. *(i) Every graph is uniquely expressible as the disjoint union of connected subgraphs, namely, its components.*

(ii) The components of a category (category-with-involution, groupoid) are themselves categories (categories-with-involution, groupoids).

(iii) Any graph-map $X \to Y$ sends each component of X into some component of Y. The image of a connected graph is connected.

(iv) A groupoid (or even a category-with-involution) A is connected if and only if A_{ij} is non-empty for all $i, j \in V(A)$.

(v) The components of a unicursal groupoid are simplicial groupoids. ■

The functors $\vec{P}$ and P give easy characterisations of the subcategory or subgroupoid generated by a subgraph of a category or groupoid (cf. p.9, Exercise 2). We note that every groupoid is a category-with-involution in a natural way, the involution being defined by $\bar{x} = x^{-1}$. It follows (Proposition 5′) that every graph-map $\theta : X \to A$, where A is a groupoid, induces a unique map $\theta^* : P(X) \to A$ satisfying (i) θ^* is a morphism of categories-with-involution, and (ii) $\theta = \sigma\theta^*$, where σ is the canonical embedding of X in $P(X)$. If we identify X with its image in $P(X)$ these conditions can be replaced by (i) θ^* is a category-map, (ii) $x\theta^* = x\theta$, $\bar{x}\theta^* = (x\theta)^{-1}$ for x an edge of X. The path $(\xi_1, \xi_2, \ldots, \xi_n)$ where ξ_i stands for x_i or $\bar{x}_i$, maps to the element $a_1 a_2 \ldots a_n$ of A, where $a_i = x_i\theta$ if $\xi_i = x_i$, and $a_i = (x_i\theta)^{-1}$ if $\xi_i = \bar{x}_i$.

PROPOSITION 7. *Let X be a subgroup of the category (groupoid) A. Let B be the subcategory (subgroupoid) generated by X. Let $\theta^* : \vec{P}(X) \to A$ (resp. $\theta^* : P(X) \to A$) be the map induced by the inclusion $X \subset A$. Then $\mathrm{Im}\,(\theta^*) = B$.*

Proof. The inclusion $X \subset B$ induces a map $\vec{P}(X) \to B$ (or $P(X) \to B$) which agrees with θ^*, so $\mathrm{Im}(\theta^*) \subset B$. Since $\mathrm{Im}(\theta^*) \supset X$ it is enough therefore to show that $\mathrm{Im}(\theta^*)$ is a subcategory (or subgroupoid). In the case of categories this follows from Proposition 1, since the vertices of $\vec{P}(X)$ are the same as the vertices of X. In the case of groupoids, $\mathrm{Im}(\theta^*)$ is a subcategory-with-involution, and since the involution in A is inversion, this means that $\mathrm{Im}(\theta^*)$ is a subgroupoid. ■

Note. This proposition tells us that the subcategory (subgroupoid) generated by X has the same vertex set as X and that its edges are all products of edges of X (and inverses of edges of X in the case of groupoids). But we must include empty products, one at each vertex of X.

We say that the subgraph X of a graph Y *spans* Y if, whenever there is a path from vertex i to vertex j in Y, there is already a path from i to j in X (in other words, $V(X) = V(Y)$ and each component of X has the same vertex set as some component of Y).

COROLLARY 1. *If the category (or groupoid) B is generated by the subgraph X then X spans B and each component of B is generated by a component of X.*

Proof. Since $\theta^* : \vec{P}(X) \to B$ (or $\theta^* : P(X) \to B$) is surjective and fixes all vertices, the components of B are precisely the images of the components of $\vec{P}(X)$ (or $P(X)$). But if X^λ $(\lambda \in \Lambda)$ are the components of X then the components of $\vec{P}(X)$ and $P(X)$ are $\vec{P}(X^\lambda)$ and $P(X^\lambda)$, respectively. These are spanned by X^λ, and the result follows immediately. ■

COROLLARY 2. *Let $\theta^\lambda : A^\lambda \to A$ $(\lambda \in \Lambda)$ be category-maps, and suppose that the A^λ are groupoids. Then the subcategory B of A generated by $\bigcup A^\lambda\theta^\lambda$ is a subgroupoid.*

Proof. The graphs $A^\lambda\theta^\lambda$, while not necessarily groupoids, are at least closed under inversion; hence $X = \bigcup A^\lambda\theta^\lambda$ is closed under inversion. By the proposition, every element of B is a product $b = x_1 x_2 \ldots x_n$ $(n \geqslant 0)$, where $x_r \in X$, so $b^{-1} = x_n^{-1} \ldots x_2^{-1} x_1^{-1}$ is also a product of edges of X and is therefore in B. ■

Exercises

1. Let X be a graph, and $\theta : X \to \mathcal{B}$ a diagram in the category $\mathcal{B}$. Let $\theta^* : \vec{P}(X) \to \mathcal{B}$ be the induced category-map. The natural definition of commutative diagram is as follows: θ is commutative if, for any pairs of vertices i, j of X, the elements of $\vec{P}_{ij}(X)$ all have the same image under θ^*. Show that this is equivalent to the definition in Ch. 2.

2. Let $C : \mathcal{G} \to \mathcal{S}$ assign to each groupoid A the set $C(A)$ whose members are the components of A. (This gives a functor by Proposition 6(iii)). Find a right adjoint for C, and show that C has no left adjoint.

3. If C is a free category, show that every full subcategory of C (in particular, every vertex semigroup) is free.

CHAPTER 4

Free groupoids

We have already constructed the free category $\vec{P}(X)$ on a graph X, and we shall now show how to construct the *free groupoid* $\pi(X)$ on X. We look for a groupoid $\pi(X)$ and a graph-map $\tau : X \to \pi(X)$ such that for every graph-map $\theta : X \to A$, where A is a groupoid, there is a unique groupoid-map $\theta^* : \pi(X) \to A$ with $\tau\theta^* = \theta$. This is our definition of free groupoid, and it determines $\pi(X)$ up to isomorphism of groupoids. Equivalently (by Proposition 3) we look for a functor $\pi : \mathfrak{D} \to \mathfrak{G}$ which is left adjoint to the forgetful functor $\mathfrak{G} \to \mathfrak{D}$. The functor π, if it exists, is then determined up to natural equivalence.

Suppose that, for a given graph X, such a groupoid $\pi(X)$ and a graph-map $\tau : X \to \pi(X)$ exist. Then τ induces a map $\tau^* : P(X) \to \pi(X)$. We identify $\overline{X} = X \cup X^{\mathrm{op}}$ with its image in $P(X)$, so that τ^* can be described as the unique category-map $P(X) \to \pi(X)$ such that $x\tau^* = x\tau$ and $\overline{x}\tau^* = (x\tau)^{-1}$ for x an edge of X. Note that with this convention the path $p = (y_1, y_2, \ldots, y_n)$, where $y_\nu = x_\nu$ or $\overline{x}_\nu$, is actually the product $p = y_1 y_2 \cdots y_n$ of the paths $y_1, y_2, \ldots, y_n$ of length 1. If the path $p = y_1 y_2 \cdots y_n$ has image $t = (y_1 \tau^*) \ldots (y_n \tau^*)$ under τ^*, then the "opposite" path $\overline{p} = \overline{y}_n \cdots \overline{y}_1$ has image $(y_n \tau^*)^{-1} \ldots (y_1 \tau^*)^{-1} = t^{-1}$. Thus the path $p\overline{p}$ (which is not an identity in $P(X)$, but a path of length $2n$) maps to tt^{-1}, an identity of $\pi(X)$. We shall show that in fact $\pi(X)$ is the "biggest" homomorphic image of $P(X)$ in which opposite paths cancel in this way.

Let $p = (y_1, y_2, \dots y_n)$ be a path in X, i.e. a directed path in $Y = X \cup X^{\mathrm{op}}$. If, for some ν, $y_{\nu+1} = \overline{y_\nu}$ (or $\overline{y_{\nu+1}} = y_\nu$, which is the same thing), then $(y_1, y_2, \dots y_{\nu-1}, y_{\nu+2}, \dots y_n)$ is also a path in X, which we call a *simple reduction of* p. We write $p \sim p'$ if there is a finite sequence of paths $p = p_0, p_1, \dots, p_k = p'$ $(k \geqslant 0)$ such that for $r = 1, 2, \dots, k$, p_r is a simple reduction of p_{r-1} or vice versa. This is an equivalence relation on the paths, and we write $\langle p \rangle$ for the equivalence class containing p. Since equivalent paths have the same source and the same target we can assign these as source and target of the equivalence class, and the set of equivalence classes then acquires a graph structure with vertex set $V(X)$. If p, q are paths such that the product pq is defined in $P(X)$ and if $p \sim p'$, $q \sim q'$, then $p'q'$ is defined and $p'q' \sim pq$. Hence the equivalence classes of paths form a category if we define $\langle p \rangle \langle q \rangle = \langle pq \rangle$ whenever pq is defined in $P(X)$; the identity elements are the classes $\langle e_i \rangle$, where e_i is the empty path at the vertex i. This category is in fact a groupoid since, for any path p, $p\overline{p}$ and $\overline{p}p$ are equivalent to empty paths, so $\langle p \rangle \langle \overline{p} \rangle$ and $\langle \overline{p} \rangle \langle p \rangle$ are identity elements. We denote this groupoid by $\pi(X)$ and call it the *fundamental groupoid* of the graph X. The reason for this name is that the definition of equivalence of paths is essentially a combinatorial definition of homotopy with fixed end-points (see p.169, Exercise 5). We shall also call it the free groupoid on X, a name which we now justify.

There is a canonical graph-map $\tau : X \to \pi(X)$ given by $x\tau = \langle x \rangle$. Suppose that A is a groupoid and $\theta : X \to A$ is a graph-map. By the universal property of P, θ extends uniquely to a map of categories-with-involution $\theta_1 : P(X) \to A$; and $\theta = \mu\theta_1$, where μ is the inclusion map $X \to P(X)$.

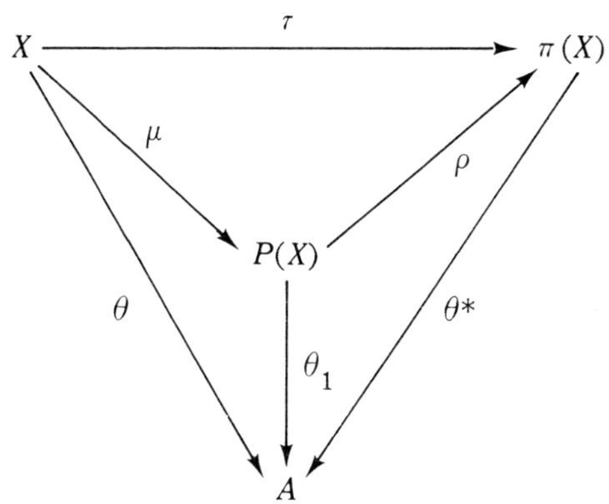

Also $\tau = \mu\rho$, where ρ is the map $p \mapsto \langle p \rangle$ from $P(X)$ to $\pi(X)$, and ρ is also a map of categories-with-involution. Now for any path p in X $p\theta_1$ and $\bar{p}\theta_1$ are inverses in A, and it follows that any two equivalent paths have the same image under θ_1. Hence $\theta_1 = \rho\theta^*$, where $\theta^* : \pi(X) \to A$ is clearly a category-map. Since $\pi(X)$ and A are groupoids, θ^* is actually a groupoid-map, and $\tau\theta^* = \mu\rho\theta^* = \mu\theta_1 = \theta$. It remains to show that θ^* is the only groupoid-map with $\tau\theta^* = \theta$. But if also $\tau\theta' = \theta$ then $\mu(\rho\theta') = \mu\theta_1$, so $\rho\theta' = \theta_1$ by the universal property of P, and clearly this implies $\theta' = \theta^*$.

This shows that $\pi(X)$ is the free groupoid on X and it follows easily that π is a functor from $\mathfrak{D}$ to $\mathfrak{G}$ (if $\theta : X \to Y$, then $\pi(\theta)$ is the groupoid map $\pi(X) \to \pi(Y)$ induced by the composite map $X \to Y \to \pi(Y)$). By Proposition 3 we therefore have:

PROPOSITION 8. *$\pi : \mathfrak{D} \to \mathfrak{G}$ is left adjoint to the forgetful functor $\mathfrak{G} \to \mathfrak{D}$.* ■

As usual, the solution of this algebraic universal problem leads immediately to a word problem: can we give an algorithm for deciding whether two paths in X have the same image in $\pi(X)$? The answer here is virtually the same as for free groups. Let $p = (y_1, y_2, \ldots, y_n)$

be a path in X, where $y_i = x_i$ or $\overline{x_i}$ (x_i an edge of X). We say that p is a *reduced path* (or reduced word) in X if, for $i = 1,2,\ldots,n\text{-}1$, $y_{i+1} \neq \overline{y_i}$, i.e. if p has no simple reductions. Clearly, every path is equivalent to at least one reduced path, and one can show, by any of the standard methods used for free groups, that no two reduced paths are equivalent. (Thus to decide whether p and q are equivalent paths or not, one need only obtain, by a finite number of simple reductions, reduced paths $p' \sim p$ and $q' \sim q$, and examine p', q' to see whether they are the same path or not). An alternative approach, often adopted in texts on group theory, is to take the set of reduced paths in X and give it the structure of a groupoid as follows. The vertices are the same as those of X; if p and q are reduced paths from i to j and from j to k, respectively, their product is defined to be the reduced word obtained from pq by successive simple reductions, (there being at most one simple reduction possible at each stage). The problem here is to show that the resulting multiplication is associative; once this is established it is easy to see that one has constructed a free groupoid on X.

We shall not give details of these arguments because, on the one hand, they differ very little from the corresponding arguments for groups and, on the other hand, we shall later solve a more general word problem (see p.73, Theorem 4). We state the result here for future reference, and warn the reader to watch for circular arguments; if he finds any, he had better do Exercise 1 below.

PROPOSITION 9. *Each equivalence class of paths in a graph X contains exactly one reduced path.*

Since the equivalence classes of paths in X form the free groupoid $\pi(X)$, we may restate this result for arbitrary free groupoids:

PROPOSITION 9′. *Let X be a graph, A a groupoid and $\tau : X \to A$ a graph-map. Then A is a free groupoid on X, with canonical map τ, if an only if*

(i) τ induces a bijection $V(X) \to V(A)$ and

(ii) each edge of A is either an identity element or is uniquely expressible as a product $(x_1\tau)^{\epsilon_1}(x_2\tau)^{\epsilon_2}\dots(x_n\tau)^{\epsilon_n}$ $(n \geqslant 1)$, where x_i is an edge of X, $\epsilon_i = \pm 1$ and if, for some $i = 1,2,\dots,n-1$, $x_i = x_{i+1}$ then $\epsilon_i = \epsilon_{i+1}$.

In particular, if A is free on X then the canonical graph map $X \to A$ is an injection. ∎

In view of the last assertion we can identify X with a subgraph of the free groupoid A, and we then say that A is freely generated by the subgraph X. The condition for this is that the non-identity elements of A should be uniquely expressible in the form $x_1^{\pm 1} x_2^{\pm 1} \dots x_n^{\pm 1}$ (x_i edges of X) with no adjacent pairs xx^{-1} or $x^{-1}x$.

Exercises

1. Let $R(X)$ denote the subgraph of $P(X)$ whose edges are all reduced paths in X (including all paths of length 0). Define a multiplication in $R(X)$ as described above and use the method of van der Waerden to show that $R(X)$ becomes a groupoid (cf. Kurosh, *Theory of groups* Vol **II** p.13). The trick here is to let M_i be the set of all reduced paths ending at the vertex i and to represent $R(X)$ as a groupoid of mappings between the sets M_i, the mappings being given by multiplication on the right in $R(X)$. Show that the representation is faithful and deduce that $R(X)$ is the free groupoid on X.

CHAPTER 5

Trees and simplicial groupoids

Having established the great similarity between free groupoids and free groups, we now point out one of the most striking differences: free groupoids can be finite! For example, if we take for X the graph $[1]$ with two vertices and one edge x joining them, then apart from the two paths of length zero, every path consists of a sequence of alternating x's and $\bar{x}$'s. Thus every path is equivalent to x or $\bar{x}$ or one of the two empty paths, and $\pi(X)$ has just two vertices and four edges. It is in fact the simplicial groupoid with two vertices, which we denote by Δ^1. Similarly $\pi([n])$ is isomorphic with Δ^n, the simplicial groupoid with $n+1$ vertices. Δ^n can thus be thought of as a free groupoid of rank n, i.e. generated freely by n edges (but note that free groupoids of the same rank are not necessarily isomorphic). More generally, we shall show that the free groupoid on any tree is simplicial and conversely that every simplicial groupoid is freely generated by any one of its maximal sub-trees. For completeness we include the necessary graph theory.

Let $p = (y_1, y_2, \ldots, y_n)$ be a path in the graph X (i.e. a directed path in $\bar{X} = X \cup X^{\mathrm{op}}$), with $y_\nu : i_{\nu-1} \to i_\nu$ for $\nu = 1,2,\ldots,n$. Here either y_ν is an edge of X or $y_\nu = \bar{x}_\nu$, where $x_\nu : i_\nu \to i_{\nu-1}$ is an edge of X. We call p a *closed path* if $i_n = i_0$. A *circuit* in X is a closed path p as above satisfying (i) $n \geqslant 1$, (ii) $i_1, i_2, \ldots, i_n$ are distinct vertices of X, and (iii) if $n = 2$ then $y_2 \neq \bar{y}_1$. In particular, any circuit is a reduced path in the sense of Ch. 4. If X has no circuits it is called *circuit-free*; a connected, circuit-free graph is called a *tree*. (Note that this should properly be called a directed tree, but our convention is that all graphs are directed). It is clear from the definition that any subgraph of a circuit-free graph is circuit-free; hence a graph is circuit-free if and only if its components are trees.

The same ideas can be approached from a different direction. If a graph X is such that any two paths between the same endpoints are equivalent in X (in the sense of Ch. 4), then we say that X is *simply-connected* (which does not imply connected). This definition is harder to work with than the definition of "circuit-free" since the equivalence of two paths in X depends on the whole of X, and it is not immediately clear, for example, that any subgraph of a simply-connected graph is simply-connected. However, the two concepts coincide, as we now show; this fact is essentially a special case of the solution of the word problem for free groupoids.

PROPOSITION 10. *A graph X is simply-connected if and only if it is circuit-free.*

Proof. Suppose that X is circuit-free. If X contains a pair of inequivalent paths p, q joining the same endpoints, we may choose such a pair of minimal total length. The closed path $p\bar{q}$ is then easily seen to be a circuit, which is impossible. Thus X is simply-connected. Conversely, if X is simply-connected, then any closed path in X is equivalent to a path of length 0 which is, of course, reduced. By Proposition 9 (the solution of the word problem), no two distinct reduced paths are equivalent, so every reduced closed path in X must have length 0. In particular, X contains no circuits. ■

COROLLARY 1. *$\pi(X)$ is unicursal if and only if X is circuit-free; $\pi(X)$ is simplicial if and only if X is a tree.*

Proof. The edges of $\pi(X)$ are the equivalence classes of paths in X, so $\pi(X)$ is unicursal if and only if X is simply-connected, and the first assertion follows. Obviously, $\pi(X)$ is connected if and only if X is connected. ∎

COROLLARY 2. *Let A be a groupoid and let X be a circuit-free subgraph of A. Then the subgroupoid of A generated by X is freely generated by it and is unicursal.*

Proof. Let B be the subgroupoid generated by X. The inclusion map $X \to B$ induces a groupoid-map $\phi: \pi(X) \to B$ which is surjective (see Proposition 7). Also, ϕ fixes all vertices of $\pi(X)$, and since $\pi(X)$ is unicursal, it is clear that ϕ must be injective. Hence ϕ is an isomorphism of groupoids. ∎

The following lemma is intuitively obvious, but requires some justification.

LEMMA. *Let X be a graph, x an edge of X, and X^* the result of removing x from X (but not removing any vertices). Then X is a tree if and only if (i) X^* is the disjoint union of two trees and (ii) the two ends of x lie one in each component of X^*.*

Proof. Suppose first that X is a tree. Then X^* is circuit-free and its components are trees. Let $x \in X_{ij}$. Then i and j lie in different components of X^*, for if p is a path in X^* from i to j of minimal length, it is easy to see that $p\bar{x}$ is a circuit in X. On the other hand, if k is an arbitrary vertex of X, there is a path in X from k to i and one from k to j. From amongst all such paths choose q of minimal length. If $q = q_1 x q_2$ or $q = q_1 \bar{x} q_2$ then q_1 is a shorter path from k to

i or j. Hence q does not contain x or $\bar{x}$, so is a path in X^*. It follows that X^* has exactly two components.

Conversely, suppose that (i) and (ii) hold. Then certainly X is connected, and we have to show that it is circuit-free. Any circuit p in X must contain x or $\bar{x}$, so we may assume that the first term of p is x or $\bar{x}$. By definition of "circuit", p contains only one term x or $\bar{x}$, and if we delete this we obtain a path in X^* connecting i and j, which is impossible. ■

THEOREM 1. *(i) Every circuit-free subgraph of a graph X is contained in a maximal circuit-free subgraph of X.*

(ii) A circuit-free subgraph of X is maximal (among all circuit-free subgraphs) if and only if it spans X.

Proof. (i) Let $\{Y^\lambda\}$ $(\lambda \in \Lambda)$ be a chain, with respect to inclusion, of circuit-free subgraphs of X. If p is a circuit in $Y = \bigcup Y^\lambda$ then, since p involves only a finite number of edges, it lies in some Y^λ, a contradiction. Thus Y is circuit-free and the result follows from Zorn's lemma.

(ii) Suppose that $Y \subset X$ is circuit-free and spans X. Since Y contains all vertices of X, any strictly larger subgraph contains an edge x not in Y. Since Y spans X the ends of x lie in the same component Y_0 of Y. By the lemma, Y_0 with x adjoined contains a circuit, so Y is a maximal circuit-free subgraph.

On the other hand, suppose that $Y \subset X$ is circuit-free and does not span X. Then there is an edge x of X whose ends lie in different components Y_1, Y_2 of Y (otherwise any path in X would lie in a component of Y). By the lemma $Y_1 \cup Y_2$ with x adjoined is a tree, so Y with x adjoined is circuit-free and Y is not maximal. ■

COROLLARY 1. *Every connected graph is spanned by a tree.* ■

COROLLARY 2. *Let A be a unicursal groupoid. Then A is free, and the subgraph X of A generates it freely if and only if X is a maximal circuit-free subgraph of A. If A is simplicial then it is generated freely by X if and only if X is a tree spanning A.*

Proof. If A is unicursal and is freely generated by X then X spans A and is circuit-free (Propositions 7 (Cor. 1) and 10 (Cor. 1)). Hence X is a maximal circuit-free subgraph of A. Conversely, if A is unicursal and X is any maximal circuit-free subgraph, then X spans A and generates freely a unicursal subgroupoid B of A (Proposition 10 (Cor. 2)). Since B spans A and A is unicursal, we must have $B = A$. The last assertion follows immediately. ■

Of course we do not need Zorn's lemma to show that the simplicial groupoid $\Delta(I)$ is free: if i is a fixed vertex, the edges (i, j), $j \neq i$, form a spanning tree and generate $\Delta(I)$ freely. However we certainly need the axiom of choice or its equivalent to prove that all unicursal groupoids are free, and the graph-theoretical form of Zorn's lemma given above is the one we need for later applications. We shall also need a standard numerical result for circuit-free graphs.

PROPOSITION 11. *Let X be a finite graph with e edges, v vertices and c components ($c = 0$ if X is empty). Then X is circuit-free if and only if $e - v + c = 0$.*

Proof. Suppose that X is circuit-free. We use induction on e to show that $e - v + c = 0$. If $e = 0$ then the vertices are all in distinct components, so $c = v$ as required. If $e > 0$, let x be any edge. Let X_0 be the component of X containing x and let X^*, X_0^* be the result of deleting x from X, X_0 respectively. Then X_0 is a tree, so by the lemma, X_0^* has two components. All other components of X are components of X^*. Thus X^* has $c^* = c+1$ components, $v^* = v$ vertices and $e^* = e-1$ edges. By induction hypothesis $e^* - v^* + c^* = 0$ since X^* is circuit-free. Hence $e - v + c = 0$.

Conversely, suppose that $e - v + c = 0$. Let X' be a maximal circuit-free subgraph of X, with parameters e', v', c'. By Theorem 1, X' spans X, so $v' = v$ and $c' = c$. But $e' - v' + c' = 0$, so $e' = e$ and $X' = X$. Thus X is circuit-free. ■

COROLLARY. *A circuit-free finite graph is a tree if and only if* $e = v - 1$. *A connected finite graph is a tree if and only if* $e = v - 1$. ■

Exercises

1. Show that $e - v + c \geqslant 0$ for any finite graph.

2. Call the edge x of the graph X *critical* if, when x is deleted, the resulting graph X^* does not span X. Prove that X is circuit-free if and only if every edge of X is critical.

3. Show that if X is circuit-free then $\vec{P}(X)$ is a unicursal category with no invertible elements except the identities (i.e. $\vec{P}(X)$ is an order relation on $I = V(X)$). The converse is false: see the next exercise.

4. Let $(I, \leqslant)$ be a pre-ordered set, and let C be the corresponding category (with vertex set I and edges all (i, j) with $i \leqslant j$). Show that C is a free category if and only if $(I, \leqslant)$ satisfies the following condition: for all $i, j \in I$ the interval $[i, j] = \{k \mid i \leqslant k \leqslant j\}$ is finite and linearly ordered.

CHAPTER 6

Fundamental groupoids of topological spaces

Let T be a topological space and I a subspace of T. We shall describe a groupoid $\pi(T,I)$ with vertex set I and edges the homotopy classes (with fixed end-points) of paths in T. A *path* in T of length r from i to j is a continuous map p from the real closed interval $[0,r]$ to T sending 0 to i and r to j. If q is a path of length s from j to k then the map

$$\begin{cases} t \mapsto (t)p & (0 \leqslant t \leqslant r) \\ t \mapsto (t-r)q & (r \leqslant t \leqslant r+s) \end{cases}$$

is a path of length $r+s$ from i to k, which we denote by $p \cdot q$. This partial multiplication of paths is associative and has identities (the paths of zero length). We therefore obtain a category $P(T)$ with vertex set T. The full subcategory with vertex set $I \subset T$ is denoted by $P(T,I)$; its edges are paths in T with ends in I. This category has a natural involution: if p is as above then $\bar{p}$ defined by $(t)\bar{p} = (r-t)p$ is a path of length r from j to i, and clearly $\overline{p \cdot q} = \bar{q} \cdot \bar{p}$, $\bar{\bar{p}} = p$. The constant paths (mapping $[0,r]$ to a point) are fixed under this involution.

If p, p' are paths of length r from i to j, write $p \simeq p'$ if p and p' are homotopic (rel. end-points), that is, if there is a continuous map h from the rectangle $[0,r] \times [0,1]$ to T such that

$$\begin{cases}(t,0)h = (t)p, & (t,1)h = (t)p' \ (0 \leqslant t \leqslant r) \\ (0,\lambda)h = i, & (r,\lambda)h = j \ (0 \leqslant \lambda \leqslant 1)\end{cases}$$

For arbitrary paths p,p' from i to j, write $p \simeq p'$ and say that p,p' are *equivalent paths*, if there exist constant paths c,c' at j such that $p \cdot c \simeq p' \cdot c'$ in the above sense. It is boring (but not difficult) to verify the following facts:

(i) the definition of $\simeq$ is unambiguous for paths of equal length;

(ii) $\simeq$ is an equivalence relation on the set of paths from i to j;

(iii) if $p \simeq p'$ $(i \to j)$ and $q \simeq q'$ $(j \to k)$ then $p \cdot q \simeq p' \cdot q'$ $(i \to k)$.

For details, see Brown [6]. It follows that the equivalence classes $\langle p \rangle$ of paths in $P(T,I)$ form a category $\pi(T,I)$ with vertex set I and multiplication $\langle p \rangle \cdot \langle q \rangle = \langle p \cdot q \rangle$. The identity element at i is the class containing the path of zero length at i; it contains also all constant paths at i. Finally, if p is a path of length r from i to j then the function

$$(t,\lambda)h = \begin{cases}(t\lambda)p & (0 \leqslant t \leqslant r,\ 0 \leqslant \lambda \leqslant 1) \\ ((2r - t)\lambda)p & (r \leqslant t \leqslant 2r,\ 0 \leqslant \lambda \leqslant 1)\end{cases}$$

is a homotopy $c \simeq p \cdot \bar{p}$, where c is the constant path of length $2r$ at i. Thus $\langle p \rangle \cdot \langle \bar{p} \rangle = \langle c \rangle$ is the identity class at i, and since $\bar{\bar{p}} = p$, $\pi(T,I)$ is a groupoid with $\langle p \rangle^{-1} = \langle \bar{p} \rangle$. The groupoid $\pi(T) = \pi(T,T)$ is the *fundamental groupoid* of T. At the other extreme, if I is a point i, we get the *fundamental group* $\pi(T,i)$ of T at the point i; these fundamental groups are just the vertex groups of $\pi(T)$.

In a path-connected space the fundamental groups at different points are isomorphic (but not canonically isomorphic). The proof is a groupoid argument and the result can be stated as

PROPOSITION 12. *All vertex groups of a connected groupoid G are isomorphic.*

Proof. Let i, j be arbitrary vertices of G. Then $\exists\, g \in G_{ij}$, and the map $x \mapsto g^{-1}xg$ is a group isomorphism $G_{ii} \to G_{jj}$. ■

The constructions P and π have obvious functorial properties. If $\alpha : X \to Y$ is a continuous map then composition with α sends paths in X to paths in Y, and equivalent paths in X go to equivalent paths in Y. If $I \subset X$, $J \subset Y$ and $I\alpha \subset J$ we therefore obtain a category-map $P(\alpha) : P(X,I) \to P(Y,J)$ and a groupoid-map $\pi(\alpha) : \pi(X,I) \to \pi(Y,J)$. If $\mathfrak{T}'$ denotes the Category whose objects are pairs (X,I) of spaces with $I \subset X$ and whose morphisms $\alpha : (X,I) \to (Y,J)$ are continuous maps $\alpha : X \to Y$ with $I\alpha \subset J$, then we have defined functors $P : \mathfrak{T}' \to \mathfrak{C}$ and $\pi : \mathfrak{T}' \to \mathfrak{G}$. By restriction we also have functors $P : \mathfrak{T} \to \mathfrak{C}$, $\pi : \mathfrak{T} \to \mathfrak{G}$ and $P : \mathfrak{T}_1 \to \mathfrak{C}_1$, $\pi : \mathfrak{T}_1 \to \mathfrak{G}_1$, where $\mathfrak{T}$ is the Category of topological spaces and $\mathfrak{T}_1$ is the Category of topological spaces with base-point.

PROPOSITION 13. *If $\alpha, \beta : X \to Y$ are homotopic, and $I\alpha \subset J$, $I\beta \subset J$, then the groupoid-maps $\pi(\alpha), \pi(\beta) : \pi(X,I) \to \pi(Y,J)$ are naturally equivalent.*

Proof. Let H be a homotopy from α to β, that is, a continuous map $H : X \times [0,1] \to Y$ such that $(x, 0)H = x\alpha$, $(x, 1)H = x\beta$. Then, for each $x \in X$, the map $p_x : t \mapsto (x, t)H$ is a path of length 1 in Y from $x\alpha$ to $x\beta$. In particular, for $i \in I$, p_i is a path from $i\alpha$ to $i\beta$, so is an edge of $P(Y, J)$. Now let q be any path in X, of length r, from i to j, where $i, j \in I$. Then the paths $(q\alpha)\cdot p_j$ and $p_i \cdot (q\beta)$ from $i\alpha$ to $j\beta$ are equivalent by means of the function h defined, for $0 \leqslant t \leqslant r + 1$, $0 \leqslant \lambda \leqslant 1$, as follows:

$$(t, \lambda)h = \begin{cases} (i, t)H & \text{if} \quad t - \lambda \leqslant 0 \\ ((t-\lambda)q, \lambda)H & \text{if} \quad 0 \leqslant t - \lambda \leqslant r \\ (j, t - r)H & \text{if} \quad t - \lambda \geqslant r. \end{cases}$$

Hence, for any edge $\langle q \rangle : i \to j$ in $\pi(X,I)$ we have, in $\pi(Y,J)$, the relation $(\langle q \rangle \pi(\alpha)) \cdot \langle p_j \rangle = \langle p_i \rangle \cdot (\langle q \rangle \pi(\beta))$, and this says that the family $\{\langle p_i \rangle\}_{i \in I}$ is a natural transformation from $\pi(\alpha)$ to $\pi(\beta)$. Since the $\langle p_i \rangle$ are invertible, it is a natural equivalence. ■

This proposition suggests some useful analogies :

(i) We shall sometimes speak of a natural equivalence between groupoid-maps $\theta, \phi : A \to B$ as a *homotopy*. Such a homotopy consists of a family $\{b_i\}$ of edges of B (one for each vertex of A) such that, for $x \in A_{ij}$, $(x\theta)b_j = b_i(x\phi)$, that is, $x\phi = b_i^{-1}(x\theta)b_j$. We write $\theta \simeq \phi$.

(ii) A groupoid-map $\alpha : A \to A$ which is homotopic to the identity map on A is of the form $x \mapsto a_i^{-1} x a_j$ $(x \in A_{ij})$, where a_i is an arbitrarily chosen edge with source i, for each i. Any such choice gives a groupoid-map $A \to A$ since if $x \in A_{ij}$, $y \in A_{jk}$ then $(a_i^{-1} x a_j)(a_j^{-1} y a_k) = a_i^{-1}(xy)a_k$. We call such a map a *deformation* of A and picture it as moving the vertex i along the edge a_i and transforming edges accordingly. The same deformation can result from different choices of the a_i. For example, if A is a group then a deformation of A is just an inner automorphism, and the same inner automorphism may be induced by different elements of the group.

(iii) A special type of deformation will play an important role later on. Let $\alpha : A \to A$ be a deformation. Then $A\alpha$ is a subgroupoid of A; for if a product $a_i^{-1} x a_j a_k^{-1} y a_l$ is defined $(x \in A_{ij}, y \in A_{kl}, a_i$ as above) then it is the image under α of the element $x a_j a_k^{-1} y$ of A_{il}. In fact $A\alpha$ is a full subgroupoid since if z has the same source and target as $x\alpha = a_i^{-1} x a_j$ then $z = (a_i z a_j^{-1})\alpha$. Write $B = A\alpha$ and let ρ be the corresponding map from A to B. If ρ restricted to B is 1_B (or equivalently $\alpha^2 = \alpha$) we say that ρ is a *deformation retraction*. Thus a groupoid-map $\rho : A \to B$, where B is a subgroupoid of A with inclusion map $\mu : B \to A$, is a deformation retraction if and only if $\mu\rho = 1_B$ and $\rho\mu \simeq 1_A$. The name is clumsy and we shall usually abbreviate it to *retraction* since we do not need this word for any other purpose. We also call B a *retract* of A if there is a retraction from A to B; this implies, of course, that A and B are equivalent groupoids.

THEOREM 2. *Let A be a groupoid and B any full subgroupoid of A which meets each component of A. Then B is a retract of A.*

Proof. It is enough to take A connected and B any non-empty full subgroupoid. Let A have vertex set I, and B have vertex set $J \subset I$, $J \neq \emptyset$. The identity map on J can be extended to a map $\sigma : I \to J$, and since A is connected we can choose $a_i \in A_{i,i\sigma}$ for each $i \in I$, with $a_i = e_i$ if $i \in J$. The resulting deformation $\alpha : x \mapsto a_i^{-1} x\, a_j$ $(x \in A_{ij})$ maps A to B and its restriction to B is 1_B. ■

COROLLARY 1. *Every vertex group of a connected groupoid A is a retract of A. Hence every connected groupoid is equivalent to a group.* ■

COROLLARY 2. *Two groupoids G and H are equivalent if and only if there is a one-one correspondence between their components such that corresponding components G^λ, H^λ have isomorphic vertex groups.*

Proof. If such a correspondence exists then G and H can be retracted to groupoids G', H' consisting of one vertex group from each component of G, H respectively. Since G', H' are isomorphic, G and H are equivalent. Conversely, if G, H are equivalent by maps $\alpha : G \to H$ and $\beta : H \to G$, then $\alpha\beta$ and $\beta\alpha$ are deformations and therefore map each component of G or H into itself. It follows that α and

β set up a correspondence $G^\lambda \Leftrightarrow H^\lambda$ between the components of G and H such that $G^\lambda\alpha \subset H^\lambda$, $H^\lambda\beta \subset G^\lambda$, and $G^\lambda \simeq H^\lambda$. Each of G^λ, H^λ is equivalent to one of its vertex groups, so it remains to show that equivalent groups are isomorphic. But this is clear since a deformation of a group is an (inner) automorphism. ■

Exercises

1. Define deformation retractions for categories exactly as for groupoids. A *skeleton* of a category A is a full subcategory containing exactly one vertex of each component of the groupoid of isomorphisms of A. Show that every skeleton of A is a retract of A, and that two categories are equivalent if and only if they possess isomorphic skeletons.

2. Let A be a groupoid and $\rho : A \to A$ a groupoid-map with $\rho^2 = \rho$. Show that ρ induces a (deformation) retraction $A \to A\rho$ if and only if it maps each A_{ij} bijectively to some A_{kl}.

3. A retraction ρ of a groupoid A onto a subgroupoid B is a *strong* retraction if it can be defined by a family $\{r_i\}$ with $r_i = e_i$ for i in B. Show that all retractions of groupoids are strong retractions.

CHAPTER 7

Limits in Categories

Throughout these notes we try to adopt the following policy. Whenever a new concept appears which is characterised by a universal property, we use this property as its definition. For example, free categories and free groupoids were defined in this way, not by the constructions of $\vec{P}(X)$ and $\pi(X)$, which are treated as existence proofs. Many other examples of such definitions will occur, and most of them are special cases of the definition of limits in arbitrary Categories. We pause here to discuss this general concept. The resulting delay in reaching the interesting parts of the theory will be compensated for in two ways. Firstly, we shall avoid some tedious repitition of standard arguments in different contexts, and secondly, the use of general theorems on limits will show clearly which parts of the theory are trivial consequences of universal properties and which depend on deeper results (such as the solution of word problems).

Let $\mathfrak{K}$ be any Category and D a non-empty directed graph. A *D-diagram* in $\mathfrak{K}$ is a graph-map $\mathbf{A} : D \to \mathfrak{K}$. If D has vertex set I and edge set E, then the diagram $\mathbf{A}$ consists of a family $\{A_i\}_{i \in I}$ of objects of $\mathfrak{K}$ and a family $\{\alpha_x\}_{x \in E}$ of $\mathfrak{K}$-morphisms, where $\alpha_x : A_i \to A_j$ if x is an edge from i to j. If $\mathbf{A}'$ is another D-diagram in $\mathfrak{K}$ with objects A' and morphisms α'_x, then a *diagram-map* or *D-map* $f : \mathbf{A} \to \mathbf{A}'$ is a family $f = \{f_i\}_{i \in I}$ of $\mathfrak{K}$-morphisms $f_i : A_i \to A'_i$

such that, for every edge x of D from i to j, $f_i a'_x = a_x f_j$. In the special case when D is a category and $\mathbf{A},\mathbf{A}'$ are functors, this is the definition of natural transformation. It is clear that all D-diagrams in $\mathfrak{K}$ are the objects of a Category $\mathfrak{K}^D$ whose morphisms are all D-maps (cf. the Category of functors and natural transformations described in Ch. 2). A *trivial* (or *constant*) D-diagram in $\mathfrak{K}$ is one in which all vertices are mapped to the same object A and all edges are mapped to the identity morphism on A. There is just one such diagram for each object A of $\mathfrak{K}$, and we denote it by $\Gamma(A)$. Also, for each $\mathfrak{K}$-morphism $\alpha : A \to B$ there is a D-map $\Gamma(\alpha) : \Gamma(A) \to \Gamma(B)$ whose components are all α. Thus we have a canonical functor $\Gamma : \mathfrak{K} \to \mathfrak{K}^D$.

Suppose now that we are given a D-diagram $\mathbf{A}$ in $\mathfrak{K}$, an object L of $\mathfrak{K}$, and a D-map $f : \mathbf{A} \to \Gamma(L)$. We say that f is a *right limit* of the diagram $\mathbf{A}$ if it has the following universal property: for every object K of $\mathfrak{K}$ and every diagram-map $g : \mathbf{A} \to \Gamma(K)$ there is a unique $\mathfrak{K}$-morphism $\beta : L \to K$ such that $g = f\Gamma(\beta)$ (i.e. $g_i = f_i\beta$ for all $i \in I$). As usual, if L and f exist for a given diagram they are uniquely determined up to $\mathfrak{K}$-isomorphism of L, and we write $L = \varinjlim \mathbf{A}$. By abuse of language we also call L the right limit of $\mathbf{A}$. Dually, if $f : \Gamma(L) \to \mathbf{A}$ is a diagram-map such that every diagram-map $g : \Gamma(K) \to \mathbf{A}$ is uniquely of the form $g = \Gamma(\beta)f$, we say that f is a *left limit* of $\mathbf{A}$ and write $L = \varprojlim \mathbf{A}$. (Freyd [**13**] uses the terms *right root* and *left root*. Mitchell [**22**] calls them *colimit* and *limit*, respectively).

If the Category $\mathfrak{K}$ is such that, for given D, every D-diagram in $\mathfrak{K}$ has a right limit, we say that $\mathfrak{K}$ *admits right limits over D*. In this case we have a functor $\varinjlim : \mathfrak{K}^D \to \mathfrak{K}$ since if $h : \mathbf{A} \to \mathbf{B}$ is a diagram-map then the composite diagram-map $\mathbf{A} \to \mathbf{B} \to \Gamma(\varinjlim \mathbf{B})$ induces a unique $\mathfrak{K}$-map $\varinjlim h : \varinjlim \mathbf{A} \to \varinjlim \mathbf{B}$. Similarly, if $\mathfrak{K}$ admits left limits over D, then $\varprojlim$ is a functor from $\mathcal{K}^D \to \mathcal{K}$. The universal properties of limits can then be reinterpreted in terms of adjoint functors as in Proposition 3 to give:

PROPOSITION 14. *If $\mathcal{K}$ admits right limits over D then the functor $\varinjlim : \mathcal{K}^D \to \mathcal{K}$ is left adjoint to the canonical functor $\Gamma : \mathcal{K} \to \mathcal{K}^D$. if $\mathcal{K}$ admits left limits over D then $\varprojlim : \mathcal{K}^D \to \mathcal{K}$ is right adjoint to Γ.* ■

Example 1. Let D have vertex set I and no edges. Then a D-diagram in $\mathcal{K}$ is just a family $\mathbf{A} = \{A_i\}_{i \in I}$ of objects of $\mathcal{K}$. If $\varprojlim \mathbf{A}$ exists it is called the *product* in $\mathcal{K}$ of the objects A_i and is denoted by $\prod = \prod_{i \in I} A_i$. The canonical maps $\pi_i : \prod \to A_i$ are called *projections*, and the universal property says there is a one-one correspondence between families of maps $\theta_i : B \to A_i$ $(i \in I)$ and maps $\theta : B \to \prod$, given by $\theta_i = \theta\pi_i$. For a finite set I we write $\prod = A_1 \times A_2 \times \ldots \times A_n$. Similarly, $\varinjlim \mathbf{A}$, if it exists, is called the *coproduct* in $\mathcal{K}$ of the objects A_i and is denoted by $\coprod_{i \in I} A_i$ (some authors write $\sum_{i \in I} A_i$). For finite coproducts we write $A_1 \amalg A_2 \amalg \ldots \amalg A_n$.

Example 2. Let D be the graph $\bullet \longrightarrow \bullet \longleftarrow \bullet$. Then the left limit L of a D-diagram $A_1 \xrightarrow{\alpha_1} A_0 \xleftarrow{\alpha_2} A_2$ (if it exists) is called the *pull-back* of α_1 and α_2 (or the fibred product of A_1 and A_2 over A_0. The diagram

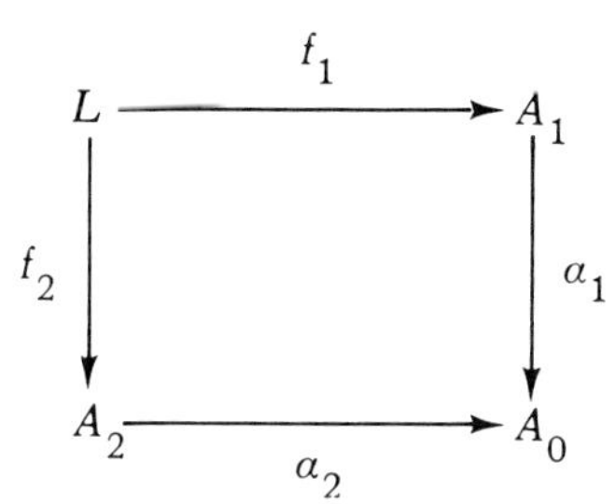

is then called a pull-back square in $\mathcal{K}$. It is commutative and has the universal property : for any $\mathcal{K}$-maps $g_i : K \to A_i$ $(i = 1,2)$ such that $g_1 \alpha_1 = g_2 \alpha_2$, there is a unique $\mathcal{K}$-map $\gamma : K \to L$ such that $g_1 = \gamma f_1$, $g_2 = \gamma f_2$. Of course the *right limit* of the diagram $A_1 \to A_0 \leftarrow A_2$ always exists and is equal to A_0.

Example 3. The right limit of a diagram $A_1 \xleftarrow{\alpha_1} A_0 \xrightarrow{\alpha_2} A_2$ is called the *push-out* of α_1 and α_2 (or the fibred coproduct), and we speak of push-out squares in the same way.

Example 4. Let D be the graph $\bullet \rightrightarrows \bullet$. Then the left limit of a D-diagram $A \underset{\alpha_2}{\overset{\alpha_1}{\rightrightarrows}} B$ is the *difference kernel* of α_1 and α_2, and the right limit is their *difference cokernel*, (also called the equaliser and coequaliser of α_1 and α_2).

If $F : \mathcal{K} \to \mathcal{L}$ is a functor and D is a graph, then from any D-diagram $\mathbf{A}$ in $\mathcal{K}$ we can obtain a D-diagram $F(\mathbf{A})$ in $\mathcal{L}$ by composing the maps $D \to \mathcal{K} \to \mathcal{L}$. In fact F induces a functor $\mathcal{K}^D \to \mathcal{L}^D$ in this way. We say that F *preserves right limits over* D if, whenever the D-diagram $\mathbf{A}$ in $\mathcal{K}$ has a right limit, the diagram $F(\mathbf{A})$ has a right limit in $\mathcal{L}$ and $\underrightarrow{\lim}\, F(\mathbf{A}) = F(\underrightarrow{\lim}\, \mathbf{A})$. (Strictly, if the $\mathcal{K}$-maps $f_i : A_i \to L$ give a right limit for $\mathbf{A}$ then the $\mathcal{L}$-maps $F(f_i)$ give a right limit for $F(\mathbf{A})$). If this is true for arbitrary graphs D we say that F *preserves right limits.* Similar definitions apply to left limits.

PROPOSITION 15. *Let $F : \mathcal{K} \to \mathcal{L}$ and $G : \mathcal{L} \to \mathcal{K}$ be functors such that (F,G) is an adjoint pair. Then F preserves right limits and G preserves left limits.*

Proof. Suppose that $\mathbf{A}$ is a D-diagram in $\mathcal{K}$ with a right limit given by the $\mathcal{K}$-maps $f_i : A_i \to A$. Applying F we obtain $\mathcal{L}$-maps $F(f_i) : F(A_i) \to F(A)$ which form a diagram-map

$F(f) : F(\mathbf{A}) \to \Gamma(F(\mathbf{A})) = F(\Gamma(A))$, and we have to verify the universal property. Let B be any object of $\mathfrak{L}$ and $g : F(\mathbf{A}) \to \Gamma(B)$ any diagram-map, consisting of $\mathfrak{L}$-maps $g_i : F(A_i) \to B$. By the adjoint property of F and G, we have a natural isomorphism $\mathfrak{K}(A, F(B)) \cong \mathfrak{L}(F(A), B)$, where A, B denote arbitrary objects of $\mathfrak{K}, \mathfrak{L}$, respectively. Applying this isomorphism to the maps g_i we obtain $\mathfrak{K}$-maps $g_i^* : A_i \to G(B)$ which, by naturality, form a map of diagrams. Since $A = \varinjlim \mathbf{A}$, there is a unique $\mathfrak{K}$-map $\gamma^* : A \to G(B)$ such that $g_i^* = f_i \gamma^*$ for all i. Going back to $\mathfrak{L}$ by the natural isomorphism we deduce that there is a unique $\mathfrak{L}$-morphism $\gamma : F(A) \to B$ such that $g_i = F(f_i)\gamma$ for all i. This shows that $F(A) = \varinjlim F(\mathbf{A})$ and the other assertion is proved by a dual argument. ■

Note. If $\mathfrak{K}$ and $\mathfrak{L}$ admit right limits over D then one half of Proposition 15 says that $F \circ (\varinjlim)$ and $(\varinjlim) \circ F^D$ are equivalent functors from $\mathfrak{K}^D$ to $\mathfrak{L}$, where F^D is the functor from $\mathfrak{K}^D$ to $\mathfrak{L}^D$ induced by F. This is a special case of Proposition 4 (ii) since if (F, G) is an adjoint pair, so is (F^D, G^D), and clearly the functors $\Gamma \circ G$ and $G^D \circ \Gamma$ from $\mathfrak{L}$ to $\mathfrak{K}^D$ are equivalent.

If the Category $\mathfrak{K}$ admits right (left) limits over arbitrary graphs D, we say simply that $\mathfrak{K}$ *admits right (left) limits* or that $\mathfrak{K}$ is *right (left) complete*. Some set-theoretical restriction on the graphs D is necessary to avoid triviality, and this is implied by our conventions. The graphs D considered are small, or lie in some fixed universe, whereas the complete Category $\mathfrak{K}$ will usually be large or lie in a bigger universe. For example, the (true) statement that the Category $\mathfrak{S}$ of sets is complete (i.e. left and right complete) means that it admits limits over graphs D for which $V(D)$ and $E(D)$ are objects of $\mathfrak{S}$. One does not allow "Graphs" as big as $\mathfrak{S}$ itself. We do not wish to labour this point but will always have a fixed set theory $\mathfrak{S}$ in mind, and all graphs, categories and groupoids are understood to be constructed in $\mathcal{S}$. In this sense the Categories $\mathcal{D}, \mathcal{C}, \mathcal{G}, \mathcal{D}_I, \mathcal{C}_I, \mathcal{G}_I$ are all complete, as we shall show in due course. The left completeness is easy; right limits in algebra are always harder, and we postpone the proof of right completeness until we have dealt with some special cases.

The easiest method of establishing completeness is to use the following result.

PROPOSITION 16. *The Category $\mathcal{K}$ is left complete if and only if it admits products and difference kernels. It is right complete if and only if it admits coproducts and difference cokernels.*

Proof. We prove the first assertion only. Let D be any graph with vertex set I and edge set $X = \bigcup X_{ij}$ $(i, j \in I)$. Let $\mathbf{A}$ be a diagram in $\mathcal{K}$ over D with objects A_i $(i \in I)$ and morphisms α_x $(x \in X)$, where $\alpha_x : A_i \to A_j$ if $x \in X_{ij}$. For convenience, we write $s(x)$, $t(x)$ for the source and target of the edge x, so that $\alpha_x : A_{s(x)} \to A_{t(x)}$. Suppose that $\mathcal{K}$ admits products and difference kernels. Then we may form products over the set X (i.e. left limits over the graph $\dot{X}$ with one vertex for each edge of D and no edges). Let $S = \prod_{x \in X} A_{s(x)}$ and $T = \prod_{x \in X} A_{t(x)}$. Since $\prod$ is a functor from $\mathcal{K}^{\dot{X}}$ to $\mathcal{K}$ the maps $\alpha_x : A_{s(x)} \to A_{(x)}$ have a product $\alpha = \prod_{x \in X} \alpha_x : S \to T$. Now let $R = \prod_{i \in I} A_i$ and let $\pi_i : R \to A_i$ be the projections. By definition of products, the morphisms $\pi_{s(x)} : R \to A_{s(x)}$ for $x \in X$ induce a morphism $\sigma : R \to S$, and the morphisms $\pi_{t(x)} : R \to A_{t(x)}$ $(x \in X)$ induce a morphism $\tau : R \to T$. By hypothesis the two morphisms τ and $\sigma\alpha$ from R to T have a difference cokernel $f : L \to R$, and we claim that $L = \varprojlim \mathbf{A}$, i.e. the morphisms $f_i = f\pi_i : L \to A_i$ form a left limit for **A**. To see this, consider any family of morphisms $g_i : K \to A_i$ $(i \in I)$, and let g be the unique morphism $K \to R$ such that $g\pi_i = g_i$. These morphisms form a diagram map from the trivial diagram $\Gamma(K)$ to **A** if and only if $g_{s(x)}\alpha_x = g_{t(x)}$ for all $x \in X$, i.e. $g\pi_{s(x)}\alpha_x = g\pi_{t(x)}$ for all x. By definition of σ and τ, this is equivalent to $g\sigma\alpha = g\tau$. Now $f : L \to R$, being the difference kernel of $\sigma\alpha$ and τ, is universal amongst such maps $g : K \to R$, and it follows easily that $L = \varprojlim$ **A**. This shows that $\mathcal{K}$ is left complete. The converse is trivial since products and difference kernels are special cases of left limits. ■

In the Category of sets $\mathcal{S}$ the product $\prod_{i \in I} A_i$ is the ordinary Cartesian product; its members are all families $\{a_i\}_{i \in I}$ with $a \in A$. The coproduct $\coprod_{i \in I} A_i$ is the disjoint union of the sets A_i (i.e. the union of disjoint copies of the A_i). The difference kernel of maps $\alpha, \beta : S \to T$ is the subset of S on which α and β agree. The difference cokernel is the set of equivalence classes on T of the equivalence relation generated by all pairs $(s\alpha, s\beta)$ with $s \in S$. These statements are easy to verify and show that $\mathcal{S}$ is a complete Category. In general, the left limit of a diagram **A** in $\mathcal{S}$, with sets A_i and maps α_x, is the subset of $\prod A_i$ consisting of all families $\{a_i\}_{i \in I}$ such that $a_i\alpha_x = a_j$ for all $x : i \to j$. The right limit of **A** is the set of equivalence classes on $\coprod A_i$ of the equivalence relation generated by all pairs $(a, a\alpha_x)$ where $a \in A_i$ and $x : i \to j$. (Here we assume, as we always may, that the A_i are disjoint and are subsets of $\coprod A_i$).

In most algebraic Categories left limits are easily constructed using the left limits of sets as building blocks. The reason for this is that one usually has forgetful functors to the Category $\mathcal{S}$ which have left adjoints and therefore preserve left limits (Proposition 15). For example the Categories $\mathcal{D}$, $\mathcal{C}$ and $\mathcal{G}$ each have two forgetful functors V and E to the Category $\mathcal{S}$ and in each case these functors have left adjoints of a rather simple form. The left adjoints of V in the three cases are the "trivial functors" T, where $T(A)$ is the *trivial* graph, category or groupoid with vertex set A. (We define a *trivial graph* to be one with no edges, and a *trivial category or groupoid* to be one with no edges other than its identity elements). The left adjoints of E are the functors $F_{\mathcal{D}}$, $F_{\mathcal{C}}$, $F_{\mathcal{G}}$, respectively, where $F_{\mathcal{D}}(A)$, $F_{\mathcal{C}}(A)$, $F_{\mathcal{G}}(A)$ are the *absolute free* graph, category and groupoid on the set A. $F_{\mathcal{D}}(A)$ consists of the disjoint union of copies of the graph  , one for each element of A. $F_{\mathcal{C}}(A)$ is a similar disjoint union of categories of type , and $F_{\mathcal{C}}$(A) is a disjoint union of simplicial groupoids of type Δ^1:

 . Clearly $F_{\mathcal{C}}(A)$ is the free category on the graph $F_{\mathcal{D}}(A)$, and $F_{\mathcal{G}}(A)$ is the free groupoid on the graph $F_{\mathcal{D}}(A)$. We shall use these constructions later, but for the present we observe that their existence implies that the forgetful functors V and E preserve left limits. Hence if a diagram $\mathbf{A}$ in $\mathcal{D}$, $\mathcal{C}$ or $\mathcal{G}$ has a left limit L then $V(L) = \varprojlim V(\mathbf{A})$ and $E(L) = \varprojlim E(\mathbf{A})$ (left limits in $\mathcal{S}$). This tells us where to look for left limits in these Categories and the rest is mere verification.

PROPOSITION 17. *The Categories* $\mathcal{D}$, $\mathcal{C}$, $\mathcal{G}$, $\mathcal{D}_I$, $\mathcal{C}_I$, $\mathcal{G}_I$ *admit left limits.*

Proof. If $\{A^\lambda\}_{\lambda\in\Lambda}$ is a family of graphs then the incidence maps $\delta_1^\lambda, \delta_2^\lambda : E(A^\lambda) \to V(A^\lambda)$ induce maps $\delta_1, \delta_2 : \prod_\lambda E(A^\lambda) \to \prod_\lambda V(A^\lambda)$ which define a graph A with $E(A) = \prod E(A^\lambda)$ and $V(A) = \prod V(A^\lambda)$. It is clear that A is the product in $\mathcal{D}$ of the A^λ. The edges from vertex $i = \{i^\lambda\}$ to $j = \{j^\lambda\}$ are all families $\{x^\lambda\}$ where $x^\lambda : i^\lambda \to j^\lambda$ in A^λ. If the A^λ are categories then A is a category with $e_i = \{e_{i^\lambda}\}$ and multiplication defined by components : $\{x^\lambda\}\{y^\lambda\} = \{x^\lambda y^\lambda\}$. Since the projections $\pi^\lambda : A \to A^\lambda$ are then category maps it follows that A is the product in $\mathcal{C}$ of the A^λ. Similarly, if the A^λ are groupoids then A becomes a groupoid if we define $\{x^\lambda\}^{-1} = \{(x^\lambda)^{-1}\}$, and A is then the product in $\mathcal{G}$ of the A^λ. Again, if $a_1, a_2 : X \to Y$ are graph maps then the vertices i of X such that $ia_1 = ia_2$, and the edges x of X such that $xa_1 = xa_2$ form a subgraph K of X which is the difference kernel in $\mathcal{D}$ of a_1 and a_2. If a_1, a_2 are category-maps (groupoid-maps) then K is a subcategory (subgroupoid) of X and is the difference kernel in $\mathcal{C}$ (in $\mathcal{G}$) of a_1 and a_2. The left completeness of $\mathcal{D}$, $\mathcal{C}$ and $\mathcal{G}$ follows from Proposition 16, and the proof of that proposition shows that in all cases $\varprojlim \mathbf{A}$ can be constructed as the subgraph, subcategory or subgroupoid of $\prod A^\lambda$ consisting of all vertices and edges whose projections in the A^λ are compatible with the maps of the diagram $\mathbf{A}$.

For $\mathcal{D}_I$, $\mathcal{C}_I$, $\mathcal{G}_I$ the construction is somewhat different. Here the vertex set is fixed and only the identity map on I is allowed as vertex map. We now have a family of forgetful functors E_{ij}, where $E_{ij}(A) = A_{ij}$ and each of these has a left adjoint (which the reader may construct for himself). Accordingly, if products exist in these Categories we must find them by taking products separately on all the (i, j)-pieces. It is easy to check that if $\{A^\lambda\}_{\lambda \in \Lambda}$ is a family of I-graphs then the graph A with vertex set I and with $A_{ij} = \prod_\lambda A^\lambda_{ij}$ for all i, j is in fact the product in $\mathcal{D}_I$ of the A^λ, and also that when the A^λ are categories (groupoids) this A carries a natural category (groupoid) structure which makes it the product in $\mathcal{C}_I$ (in $\mathcal{G}_I$). As for difference kernels, if a_1, a_2 are morphisms $X \to Y$ in $\mathcal{D}_I$, their difference kernel in $\mathcal{D}$ is in $\mathcal{D}_I$ and is therefore their difference kernel in $\mathfrak{D}_I$. The same applies to $\mathfrak{C}_I$ and $\mathfrak{G}_I$, and the result follows. ■

Note. When we write $\prod A^\lambda$ or $\varprojlim \mathbf{A}$ etc. for a diagram of graphs categories or groupoids we shall always mean the product or limit in the appropriate Category $\mathfrak{D}$, $\mathfrak{C}$ or $\mathfrak{G}$ even when the A^λ all have the same vertex set. Thus $\Delta^1 \times \Delta^1$ means Δ^3 (simplicial groupoid with four vertices), not Δ^1 which is the product of Δ^1 and Δ^1 in the Category of groupoids with two vertices.

There is one other general result on limits, closely related to Proposition 15, which will be used in many special cases later. It says essentially that right limits commute with right limits (and left with left). Let $\mathfrak{K}$ be any Category and let C, D be arbitrary graphs with vertex sets I and J respectively. It is easy to see that the diagram Categories $(\mathfrak{K}^C)^D$ and $(\mathfrak{K}^D)^C$ are isomorphic. A C-diagram in $\mathfrak{K}^D$ assigns to each $i \in I$ a D-diagram consisting of a family of objects $\{A_{ij}\}_{j \in J}$ and a family of morphisms $a_{id} : A_{ij} \to A_{ij'}$, where $d : j \to j'$ is an edge of D. It assigns to each edge $c : i \to i'$ of C a D-map consisting of morphisms $a_{cj} : A_{ij} \to A_{i'j}$. Thus an object of $(\mathfrak{K}^D)^C$ consists of objects A_{ij} and morphisms a_{id}, a_{cj} (of $\mathfrak{K}$), as above, satisfying the relations $a_{id}a_{cj'} = a_{cj}\,a_{i'd}$. By transposing all pairs of suffixes we obtain the corresponding object of $(\mathfrak{K}^C)^D$. For simplicity we denote both diagrams by $\mathbf{A}$ and call $\mathbf{A}$ a (C, D)-diagram in $\mathfrak{K}$. The morphisms $\mathbf{A} \to \mathbf{B}$ of $(\mathfrak{K}^C)^D$ or $(\mathfrak{K}^D)^C$ are then all families of $\mathfrak{K}$-maps $\phi_{ij} : A_{ij} \to B_{ij}$ compatible with the maps $a_{..}$ and $\beta_{..}$ in the two diagrams $\mathbf{A}$ and $\mathbf{B}$. Thus the two Categories $(\mathfrak{K}^C)^D$ and $(\mathfrak{K}^D)^C$ can be identified with each other.

Now suppose that $\mathfrak{K}$ admits right limits. Then the canonical functor $\Gamma_D : \mathfrak{K} \to \mathfrak{K}^D$ has a left adjoint $\varinjlim_D : \mathfrak{K}^D \to \mathfrak{K}$, and these two functors induce an adjoint pair of functors $\mathfrak{K}^C \rightleftarrows (\mathfrak{K}^D)^C$. Replacing

$(\mathcal{K}^D)^C$ by $(\mathcal{K}^C)^D$, we see that the canonical functor $\Gamma_D : \mathcal{K}^C \to (\mathcal{K}^C)^D$ has a left adjoint; in other words, $\mathcal{K}^C$ admits right limits over D (and this for all graphs C, D). Furthermore, the functor $\varinjlim_C : \mathcal{K}^C \to \mathcal{K}$, being a left adjoint, preserves right limits over D, by Proposition 15. Thus, for any (C,D)-diagram $\mathbf{A}$, $\varinjlim_C (\varinjlim_D \mathbf{A}) = \varinjlim_D (\varinjlim_C \mathbf{A})$, and we have

PROPOSITION 18. *If $\mathcal{K}$ admits right limits then, for all graphs C, D, the functors $\varinjlim_C \circ \varinjlim_D$ and $\varinjlim_D \circ \varinjlim_C$ from $(\mathcal{K}^C)^D = (\mathcal{K}^D)^C$ to $\mathcal{K}$ are naturally equivalent. A similar statement is true for left limits.* ■

Note. It is not true that $\mathcal{K}^{C \times D}$ is isomorphic with $(\mathcal{K}^C)^D$ and $(\mathcal{K}^D)^C$, because of the lack of "identity edges" in the graphs. One may, without loss of generality (but with some loss of convenience) restrict C and D to be categories and consider only diagrams $C \to \mathcal{K}$ etc. which are *category-maps*. In this case the functor Categories $(\mathcal{K}^C)^D$, $(\mathcal{K}^D)^C$ and $\mathcal{K}^{C \times D}$ are all isomorphic, and the functor $\varinjlim_{C \times D}$ is also equivalent to the two composite limit functors.

Exercises

1. Show that $\mathcal{D}$ admits right limits.

2. Let α, β be the two groupoid-maps from Δ^0 to Δ^1. Show that the difference cokernel in $\mathcal{G}$ of α and β is an infinite cyclic group.

3. Show that the forgetful funtor $E : \mathcal{G} \to \mathcal{S}$ does not have a right adjoint. (Hint: show that E does not preserve right limits).

4. Prove that if T, T' are topological spaces with subspaces I, I', then $\pi(T \times T', I \times I') \cong \pi(T, I) \times \pi(T', I')$.

CHAPTER 8

Universal morphisms in $\mathcal{D}$, $\mathcal{C}$ and $\mathcal{G}$

In order to prove the existence of right limits of diagrams of graphs, categories or groupoids, it is convenient to separate the roles of the vertices and of the edges. For the vertices we know what to expect. The forgetful functor V has a right adjoint Δ in all three cases ($\Delta(I)$ is the simplicial groupoid with vertex set I, considered as a graph, a category or a groupoid according to the context). Hence V preserves right limits (Proposition 15), that is, the vertex set of a right limit of groupoids (etc.) is the right limit of the vertex sets. Given a diagram **A** of (say) groupoids A^λ with vertex sets I^λ we can start by constructing $I = \varinjlim I^\lambda$ and looking for an appropriate I-groupoid. We show now that one can pass from the given diagram to a diagram in $\mathcal{G}_I$ (where the problem is often easier) by means of "universal morphisms" $A^\lambda \to B^\lambda$ induced by the canonical maps $I^\lambda \to I$. The construction is a special case of right limit and includes such basic constructions as free categories, free groupoids and free products of groups. Furthermore, we can solve the word problem in this special case, which is not possible for general right limits.

Let $\theta : A \to B$ be a groupoid-map and write $I = V(A)$, $J = V(B)$, $\sigma = V(\theta) : I \to J$. Let $T(I)$, $T(J)$ denote the trivial groupoids with vertex sets I and J respectively. Then σ induces a groupoid-map $T(\sigma) : T(I) \to T(J)$ and we have canonical injections $T(I) \to A$, $T(J) \to B$ making the diagram

$$\begin{array}{ccc} T(I) & \xrightarrow{T(\sigma)} & T(J) \\ \downarrow & & \downarrow \\ A & \xrightarrow[\theta]{} & B \end{array}$$

commute. If this diagram is a push-out square in $\mathcal{G}$, we say that θ is a *universal groupoid-map*. Thus θ is universal if, for every groupoid-map $\phi : A \to C$ whose vertex map is of the form $\tau = \sigma\tau^*$, there is a unique groupoid-map $\phi^* : B \to C$ such that $\phi = \theta\phi^*$ and $V(\phi^*) = \tau^*$.

Suppose that we are given the groupoid A with vertex set I and the map $\sigma : I \to J$. Then, by the uniqueness of push-outs, the groupoid B and the map θ (if they exist) are determined up to isomorphism. We are therefore faced with two problems: (i) to prove the existence of universal morphisms and (ii) to find the structure of B, i.e. to solve the word problem for B. The same problems can be posed for graphs and categories, and we shall deal with these first. The definitions of universal graph-maps and universal category-maps are exactly as above with $\mathcal{G}$ replaced by $\mathcal{D}$ or $\mathcal{C}$. In $\mathcal{C}$, $T(I)$ denotes the trivial category with vertex set I, and in $\mathcal{D}$, $T(I)$ denotes the graph with vertex set I and no edges.

For graphs the solution is trivial. Let X be a graph with vertex set I and let $\sigma : I \to J$ be any map. We form a new graph X^σ as follows: the vertex set of X^σ is J, the edges of X^σ are just the edges of X, and the incidence maps $E(X) \to J$ are $\delta_1\sigma$ and $\delta_2\sigma$, where δ_1, δ_2 are the incidence maps of X. Then the identity map on $E(X)$, together with $\sigma : I \to J$, gives a graph map $\sigma^* : X \to X^\sigma$, and we leave it to the reader to check that σ^* is the required universal graph-map.

Now suppose that A is a category and $\sigma : I \to J$ a map, where $I = V(A)$. We may view A as a graph and form the graph A^σ and the canonical map $A \to A^\sigma$, which we now denote by σ_1. Let $\phi : A \to C$ be any category-map whose vertex map is of the form $\tau = \sigma\tau^*$. Then, by the universal property of σ_1 in $\mathcal{D}$, there is a unique graph-map $\phi_1 : A^\sigma \to C$ such that $\phi = \sigma_1 \phi_1$ and $V(\phi_1) = \tau^*$. Of course A^σ is not in general a category, and σ_1, ϕ_1 are not category-maps. However, the graph-map ϕ_1 from A^σ into the category C induces a category-map $\phi_2 : \vec{P}(A^\sigma) \to C$, where $\vec{P}(A^\sigma)$ is the category of directed paths in A^σ (see Ch. 3), and we have a commutative diagram

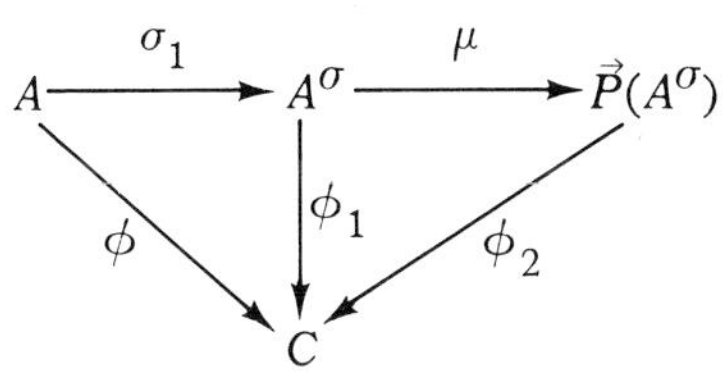

where μ is the inclusion map of A^σ in $\vec{P}(A^\sigma)$ (with the usual identification of edges and paths of length 1). Since $\vec{P}(A^\sigma)$ has the same vertex set as A^σ, namely J, the map $\sigma_2 = \sigma_1 \mu : A \to \vec{P}(A^\sigma)$ has vertex map σ and looks very like the hoped-for universal map. The only snag is that it is not a category-map since we have not yet used the category structure of A. If x is an edge of A we denote by x^σ the corresponding edge of A^σ to avoid ambiguity. If $xy = z$ in A then $x^\sigma y^\sigma$ is defined in $\vec{P}(A^\sigma)$ and is a path of length 2; it cannot be equal to z^σ which is a path of length 1. We must therefore replace $\vec{P}(A^\sigma)$ by an image category in which $x^\sigma y^\sigma$ and z^σ become equal. Also, if $i^\sigma = j$, we must equate the paths e_i^σ and e_j (of lengths 1, 0, respectively).

Let $p = x_1^\sigma x_2^\sigma \ldots x_n^\sigma$ be an edge of $\vec{P}(A^\sigma)$, that is, a path of length n in A^σ, from j to j'. If for some ν $(1 \leqslant \nu < n)$ the product $x_\nu x_{\nu+1}$

is defined in A and has the value x, then $x_1^\sigma \dots x_{\nu-1}^\sigma x^\sigma x_{\nu+2}^\sigma \dots x_n^\sigma$ is a path from j to j'. Also if some x_ν $(1 \leqslant \nu \leqslant n)$ is an identity element of A then $x_1^\sigma \dots x_{\nu-1}^\sigma x_{\nu+1}^\sigma \dots x_n^\sigma$ is a path from j to j'. (If $n = 1$ this is the empty path at j). We call these two types of modified paths *elementary reductions* of p, and we write $p \sim p'$ if there exist paths $p = p_1, p_2, \dots, p_k = p'$ such that, for each $r = 1, 2, \dots, k\text{-}1$, p_{r+1} is an elementary reduction of p_r, or *vice versa*. This is an equivalence relation on the edges of $\vec{P}(A^\sigma)$, and equivalent paths have the same source and the same target. Thus the equivalence classes $[p]$ form a graph with vertex set J. We denote it by $U_\sigma(A)$. If $p \sim p'$, $q \sim q'$ and pq is defined in $\vec{P}(A^\sigma)$, then $p'q'$ is defined and $pq \sim p'q'$, so $U_\sigma(A)$ is a category with multiplication $[p][q] = [p\,q]$ and identity elements $[e_j]$ $(j \in J)$, where e_j is the empty path at j. Further, if π is the canonical map $p \mapsto [p]$ from $\vec{P}(A^\sigma)$ to $U_\sigma(A)$, then the map $\sigma_3 = \sigma_2\pi = \sigma_1\mu\pi$ from A to $U_\sigma(A)$ is a category-map; for if $xy = z$ in A then $x^\sigma y^\sigma \sim z^\sigma$, which implies $[x^\sigma][y^\sigma] = [z^\sigma]$, and if e_i is an identity of A then $e_i^\sigma \sim e_j$ $(j = i\sigma)$, which implies that e_i maps to the identity element $[e_j]$ of $U_\sigma(A)$.

Now if $xy = z$ in A, then the paths $x^\sigma y^\sigma$ and z^σ have the same image in C under the category map ϕ_2, namely $(x\phi)(y\phi) = z\phi$. Also e_i^σ has the same image $e_i\phi$ as the empty path e_j $(j = i\sigma)$. Since ϕ_2 is a category-map, it follows that any elementary reduction of a path p has the same image in C as p. Hence equivalent paths have the same image in C and there is a category-map $\phi_3 : U_\sigma(A) \to C$ such that the diagram

$$\begin{array}{ccccccc} A & \xrightarrow{\sigma_1} & A^\sigma & \xrightarrow{\mu} & \vec{P}(A^\sigma) & \xrightarrow{\pi} & U_\sigma(A) \\ & \searrow^{\phi} & \downarrow{\scriptstyle\phi_1} & & \swarrow{\scriptstyle\phi_2} & \swarrow_{\phi_3} & \\ & & & C & & & \end{array}$$

commutes. Thus the category-map $\phi : A \to C$ factorises through $U_\sigma(A)$ in the form $\phi = \sigma_3\phi_3$, where $\sigma_3 = \sigma_1\mu\pi$ and $V(\phi_3) = V(\phi_1) = \tau^*$. The map ϕ_3 is uniquely determined by ϕ since $U_\sigma(A)$ is generated by the image of A under σ_3, except possibly for some isolated identity elements whose images are determined by the vertex map τ^*. This shows that σ_3 is the universal category-map corresponding to the vertex map σ applied to A.

PROPOSITION 19. *(i) For every I-category A and every map $\sigma : I \to J$ there is a universal category-map $A \to U_\sigma(A)$ with vertex map σ.*

(ii) U_σ is a functor from $\mathcal{C}_I$ to $\mathcal{C}_J$, uniquely determined up to natural isomorphism by σ.

(iii) If $\tau : J \to K$ then $U_{\sigma\tau} = U_\tau \circ U_\sigma$. Products of universal category-maps are universal.

Proof. (i) has already been proved.

(ii) If $\theta : A \to B$ is a morphism of $\mathcal{C}_I$, then the universal category-map $B \to U_\sigma(B)$ composed with θ gives a category-map $A \to U_\sigma(B)$ whose vertex map is σ. It therefore induces a unique map $U_\sigma(\theta) : U_\sigma(A) \to U_\sigma(B)$ such that the diagram

$$\begin{array}{ccc} A & \longrightarrow & U_\sigma(A) \\ \theta \big\downarrow & & \big\downarrow U_\sigma(\theta) \\ B & \longrightarrow & U_\sigma(B) \end{array}$$

commutes. It is clear that this gives us a functor $U_\sigma : \mathcal{C}_I \to \mathcal{C}_J$. We know also that $U_\sigma(A)$ is determined up to isomorphism by A and σ, and it is trivial to check that these isomorphisms are natural.

(iii) Let $\rho = \sigma\tau : I \to K$. If $\phi : A \to C$ is any category-map whose vertex map has ρ as a left factor, then ϕ induces $\phi_1 : U_\sigma(A) \to C$.

The vertex map of ϕ_1 has τ as a left factor, so ϕ_1 induces $\phi_2 : U_\tau(U_\sigma(A)) \to C$. If follows easily that the product of the universal maps $A \to U_\sigma(A) \to U_\tau(U_\sigma(A))$ is universal with vertex map ρ, and hence that $U_\tau(U_\sigma(A)) \cong U_\rho(A)$ (natural equivalence of functors). ■

When we turn to the corresponding problem for groupoids we find an unexpected bonus. If the category A above happens to be a groupoid then $U_\sigma(A)$, being generated as a category (apart from isolated identity elements) by the image of a groupoid, is itself a groupoid (see Proposition 7, Corollary 2). The categories $T(I), T(J)$ are also groupoids and the maps in the diagram

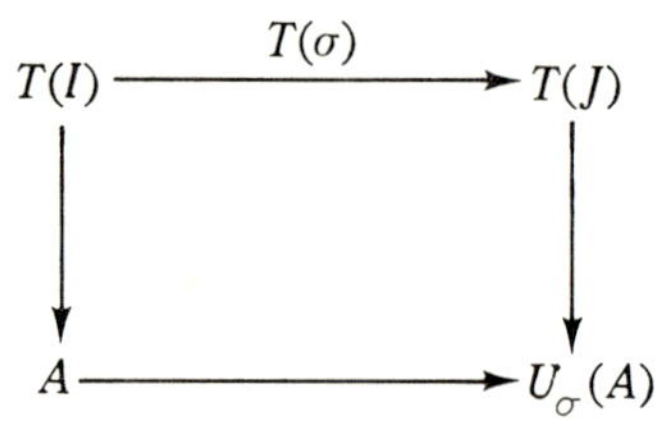

are all groupoid-maps. Since this is a push-out square in $\mathcal{C}$ it is *a fortiori* a push-out square in $\mathcal{G}$, so the universal category-map $A \to U_\sigma(A)$ is also a universal groupoid-map in this case. (We are, of couse, identifying $\mathcal{G}$ with a sub-Category of $\mathcal{C}$ by the forgetful functor). Hence we have

PROPOSITION 19′. *(i) For every I-groupoid A and every map $\sigma : I \to J$ there is a universal groupoid-map $A \to U_\sigma(A)$ with vertex map σ.*

(ii) U_σ is a functor from $\mathcal{G}_I$ to $\mathcal{G}_J$, uniquely determined up to natural isomorphism by σ. It is the restriction to $\mathcal{G}_I, \mathcal{G}_J$ of the functor U_σ of Proposition 19.

(iii) If $\tau : J \to K$, then $U_{\sigma\tau} = U_\tau \circ U_\sigma$. Products of universal groupoid-maps are universal groupoid-maps. ■

PROPOSITION 20. *Let X be a graph with vertex set I, and let σ be map from I to J. If A is the free category (free groupoid) on X then $U_\sigma(A)$ is the free category (free groupoid) on the graph X^σ.*

Proof. Check the universal property; or see the examples below.■

Example 1. Free categories. Any graph X is of the form X_0^σ, where X_0 is the result of "pulling X apart". More precisely, let $E = E(X)$ and let X_0 be the absolute free graph $F_{\mathfrak{D}}(E)$ (see p.56). X_0 can be described as the product $T(E) \times [1]$, where $[1]$ denotes the graph $\bullet \longrightarrow \bullet$, and $T(E)$ is the trivial graph with vertex set E. Or again, $X_0 = \coprod_{x \in E} [1]$, where $\coprod$ denotes coproduct in $\mathfrak{D}$, which is obviously just disjoint union. There is a unique graph-map $X_0 \to X$ sending each edge to itself and this map is a universal graph-map. Thus $X = X_0^\sigma$, where σ denotes the corresponding vertex map.

Let C_0 be the free category on X_0. Then $C_0 = F_{\mathcal{C}}(E) = \coprod_{x \in E} \{1\}$. where $\{1\}$ denotes the free category  generated by $[1]$. One sees easily that $U_\sigma(C_0)$ is the free category on X, either by checking the universal property or by noticing that the free category functor $\vec{P} : \mathfrak{D} \to \mathcal{C}$ is a left adjoint and so preserves push-outs.

Example 2. Free groupoids. The free groupoid on the graph $[1]$ is the simplicial groupoid Δ^1. The free groupoid G_0 on X_0 is therefore $G_0 = F_{\mathcal{G}}(E) = \coprod_{x \in E} \Delta^1$, and as above, $U_\sigma(G_0)$ is the free groupoid on X. Proposition 20 is now seen as an application of the relation $U_{\sigma\tau} = U_\tau \circ U_\sigma$.

Example 3. Free products of groups. Let G be a *totally disconnected* groupoid, i.e. the disjoint union of groups $G^i (i \in I)$. Let J be a one-element set and σ the unique map $I \to J$. Then $U_\sigma(G)$ is a group, and there are canonical group-maps $\gamma^i : G^i \to U_\sigma(G)$ induced by $G \to U_\sigma(G)$. The universal property of $U_\sigma(G)$ says that for any family of group-maps $\theta^i : G^i \to H$ there is a unique group-map $\theta : U_\sigma(G) \to H$ such that $\theta^i = \gamma^i\theta$ for all $i \in I$. This is precisely the statement that $U_\sigma(G)$ is the coproduct in $\mathcal{G}_1$ (the Category of groups) of the groups G^i, i.e. (definition) $U_\sigma(G)$ is the *free product* of the groups G^i. Free products of semigroups-with-1 can be treated in the same way.

Exercises

1. Given a J-groupoid B and a map $\sigma : I \to J$, let $W_\sigma(B)$ denote the graph whose vertex set is I and whose edges from i_1 to i_2 are (in one-one correspondence with) the edges of B from $i_1\sigma$ to $i_2\sigma$. Show that $W_\sigma(B)$ has a natural groupoid structure and that W_σ is a functor from $\mathcal{G}_J$ to $\mathcal{G}_I$. Show also that W_σ is a right adjoint to $U_\sigma : \mathcal{G}_I \to \mathcal{G}_J$. (Hence U_σ preserves right limits, which is easy to check directly).

2. Let

$$\begin{array}{ccc} A_0 & \xrightarrow{\theta_1} & A_1 \\ {\scriptstyle\theta_2}\downarrow & & \downarrow{\scriptstyle\phi_1} \\ A_2 & \xrightarrow[\phi_2]{} & A \end{array}$$

be a push-out square in $\mathcal{G}$. Show that if θ_1 is a universal groupoid-map, then so is ϕ_2. In particular, if A_0 and A_1 are trivial groupoids and θ_1 is arbitrary then ϕ_2 is universal. (Hint: the forgetful functor $V : \mathcal{G} \to \mathcal{S}$ preserves push-outs).

CHAPTER 9

Right limits in $\mathcal{C}$ and $\mathcal{G}$

The existence of right limits in $\mathfrak{D}$ has been set as an exercise and is easy enough. One simply takes right limits of the vertex sets and the edge sets separately and verifies that the resulting limit sets have a graph structure with the required properties. In $\mathcal{C}$ and $\mathcal{G}$ this construction fails in general (see Exercises 1,2,3, p.59) but it is clear that for coproducts it still works. Coproducts in $\mathcal{C}$ or $\mathcal{G}$ are just disjoint unions, and we use the notations $\coprod A^{\lambda}$, $A^1 \amalg A^2$, etc. According to Proposition 16 we need only show, therefore, that difference cokernels always exist in $\mathcal{C}$ and $\mathcal{G}$.

Let θ, $\phi : A \to B$ be category-maps with corresponding vertex maps $\theta_0, \phi_0 : I \to J$, and let $\sigma_0 : J \to K$ be the difference cokernel of θ_0 and ϕ_0 (in $\mathcal{S}$). Then σ_0 induces a universal category-map $\sigma : B \to U_{\sigma_0}(B)$, and if $\gamma : B \to C$ is any category-map such that $\theta\gamma = \phi\gamma$, then γ has the form $\gamma = \sigma\gamma^*$, where $\gamma^* : U_{\sigma_0}(B) \to C$ is a category-map, uniquely determined by γ. It follows immediately from the definitions that γ is the difference cokernel in $\mathcal{C}$ of θ and ϕ if and only if γ^* is the difference cokernel of $\theta' = \theta\sigma$ and $\phi' = \phi\sigma$. (θ', $\phi' : A \to B' = U_{\sigma_0}(B)$). Now if x is any edge of A, the edges $x\theta'$, $x\phi'$ of B' have the same source and the same target, and this makes the rest of the construction easy. For edges p, q of B' we write $p \sim q$ whenever $p = b_1(a\theta')b_2$, $q = b_1(a\phi')b_2$ for some $a \in A$, $b_1, b_2 \in B'$. This relation on B' generates an equivalence relation $\equiv$, and the equivalence classes form a graph D with the same vertex set K as B'. Since $p \sim q$ implies $bp \sim bq$ and $pb \sim qb$, it is clear that $p \equiv q$ implies $bp \equiv bq$ and $pb \equiv qb$. Hence $p \equiv q$ and $p' \equiv q'$ implies $pp' \equiv qp' \equiv qq'$ (whenever pp' is defined). Thus D inherits a category structure from B' and the canonical map $\pi : B' \to D$ is the difference cokernel (in $\mathcal{C}$) of θ' and ϕ'. It follows that θ, ϕ have difference cokernel $\sigma\pi : B \to D$. If A and B are groupoids, and θ, ϕ are groupoid-maps, then $B' = U_{\sigma_0}(B)$ and $D = B\pi$ are groupoids, and $\sigma\pi : B \to D$ is the difference cokernel in $\mathcal{G}$ of θ and ϕ. Thus $\mathcal{C}$ and $\mathcal{G}$ admit right limits.

For right limits in $\mathcal{D}_I$, $\mathcal{C}_I$ and $\mathcal{G}_I$ we may proceed as follows. Let $\mathbf{A}$ be a D-diagram in one of these Categories, and let $T(I)$ denote the trivial graph, category or groupoid with vertex set I. Let D' be the graph obtained by adjoining to D one new vertex 0 and new edges $x_\lambda : 0 \to \lambda$ one for each vertex λ of D. Let $\mathbf{A}'$ be the D'-diagram in which D maps as before (with $\lambda \mapsto A^\lambda$, say), 0 maps to $T(I)$, and x_λ maps to the canonical embedding $T(I) \to A^\lambda$, for each λ. Then the right limit of $\mathbf{A}'$ in $\mathcal{D}$, $\mathcal{C}$ or $\mathcal{G}$ is (with the obvious abuse of language) the right limit of $\mathbf{A}$ in $\mathcal{D}_I$, $\mathcal{C}_I$ or $\mathcal{G}_I$. In particular, we may recover right limits in the Category of groups by adjoining the trivial group to all diagrams in this way and taking limits in the Category of groupoids.

Combining these results with Proposition 17, we have now proved

THEOREM 3. *The Categories* $\mathcal{D}$, $\mathcal{C}$, $\mathcal{G}$, $\mathcal{D}_I$, $\mathcal{C}_I$, $\mathcal{G}_I$ *are complete.* ■

The presentation of categories and groupoids by generators and relations is a special case of the above construction.. Suppose that we are given a graph X and a set R whose members are ordered pairs of category-words on the alphabet X (i.e. ordered pairs of edges of the free category $\vec{P}(X)$). Let $F_{\mathcal{C}}(R)$ be the absolute free category on R. Then there are unique category-maps θ_1, $\theta_2 : F_{\mathcal{C}}(R) \to \vec{P}(X)$ given by $r\theta_1 = r_1$, $r\theta_2 = r_2$, where $r = (r_1, r_2) \in R$. If $\phi : \vec{P}(X) \to C$ is the difference cokernel (in $\mathcal{C}$) of θ_1 and θ_2 we write $C = \text{cat}\ (X; R)$ and say that C is the category with generators X and defining relations R (or defining relations $r_1 = r_2$, $(r_1, r_2) \in R$). We also say that the triple (X, R, ϕ_0), where $\phi_0 : X \to C$ is the restriction of ϕ to X, is a *presentation* of C in the Category $\mathcal{C}$. If A is any category and $\alpha : X \to A$ is a graph-map, we say that the relations R *hold in* A *under the map* α if for all $(r_1, r_2) \in R$ we have $r_1\alpha^* = r_2\alpha^*$, where $\alpha^* : \vec{P}(X) \to A$ is the category-map induced by α. The category $C = \text{cat}\ (X; R)$ and the canonical graph-map $\phi_0 : X \to C$ can then be characterised by the following properties:

(i) the relations R hold in C under ϕ_0;

(ii) if the relations R hold in A under $\alpha : X \to A$, then $\alpha = \phi_0\alpha'$ uniquely, where $\alpha' : C \to A$ is a category-map.

Note that if R is empty then so is $F_{\mathcal{C}}(R)$, and $C = \vec{P}(X)$.

Similarly, if R is a set of ordered pairs of groupoid-words on the alphabet X (i.e. edges of $\pi(X)$) we obtain a groupoid $G = \text{gpd}(X; R)$ as the difference cokernel in $\mathcal{G}$ of the groupoid-maps $\theta_1, \theta_2 : F_{\mathcal{G}}(R) \to \pi(X)$ given by $r\theta_1 = r_1$, $r\theta_2 = r_2$ $(r = (r_1, r_2) \in R)$, and all the above remarks apply (with obvious modifications).

Exercises

1. Let $C = \text{cat}\ (X; R)$, where X has vertex set I, and let $\sigma : I \to J$ be any map. Let R^σ be the set of pairs (r_1^σ, r_2^σ) $((r_1, r_2) \in R)$, where r_k^σ denotes the image of r_k in $\vec{P}(X^\sigma)$, i.e. re-interpret the words r_k as words on the alphabet X^σ. Prove that $U_\sigma(C) = \text{cat}(X^\sigma; R^\sigma)$. Prove

also that $U_\sigma(C) = \text{cat}(X; R \cup S)$, where S is the set of pairs $(e_i, e_{i'})$ such that $i\sigma = i'\sigma$.

2. Show that the forgetful functor $\mathcal{G} \to \mathcal{C}$ has both a left and a right adjoint (hence preserves left and right limits).

CHAPTER 10

The word problem for U_σ

Let A be a category with vertex set I, and let σ be a map from I to a set J. Then $U_\sigma(A)$ is a category with vertex set J and its edges, according to the construction in Ch. 8, are equivalence classes of (directed) paths in the graph A^σ. The equivalence relation was defined in terms of "elementary reductions" of two types: deleting identity elements and multiplying in A when possible. We shall say that a path in A^σ is *σ-reduced* if it has no elementary reductions. (See p.64 for the precise definitions).

We shall identify edges of A with their images in A^σ to simplify the notation. We cannot, of course, do this with vertices. A path p in A^σ is then either one of the empty paths e_j ($j \in J$) or is of the form $x_1 x_2 \dots x_n$ (strictly a sequence $(x_1, x_2, \dots x_n)$) where the x_ν are edges of A satisfying incidence relations $x_\nu\delta_2\sigma = x_{\nu+1}\delta_1\sigma$ ($\nu = 1, 2, \dots, n\text{-}1$). The σ-reduced paths are of two types:

(i) the empty paths e_j ($j \in J$);

(ii) paths $x_1 x_2 \dots x_n$ ($n \geqslant 1$), where the x_ν are *non-identity* edges of A satisfying $x_\nu\delta_2\sigma = x_{\nu+1}\delta_1\sigma$ ($\nu = 1, 2 \dots n\text{-}1$) but such that the products $x_\nu x_{\nu+1}$ are *not* defined in A, i.e. $x_\nu\delta_2 \neq x_{\nu+1}\delta_1$ ($\nu = 1, 2, \dots, n\text{-}1$).

THEOREM 4. *Each edge of $U_\sigma(A)$ is represented by exactly one σ-reduced path.*

Proof. It is clear from the definition of equivalence of paths that each equivalence class contains at least one σ-reduced path. To show that it contains at most one, we adapt the method of van der Waerden for the case of free groups.

Each σ-reduced path has a source and target in J. Let S_j be the set of all σ-reduced paths with target j. If x is an edge of A from i to i', not an identity element, and if $i\sigma = j$, $i'\sigma = j'$, we define a map $\rho_x : S_j \to S_{j'}$ as follows:

(i) if $p \in S_j$ is empty then $p\rho_x = x \in S_{j'}$;

(ii) if $p = x_1 x_2 \ldots x_n \in S_j$ $(n \geqslant 1)$ there are three cases:

(a) if the product $x_n x$ is not defined in A then $p\rho_x = x_1 x_2 \ldots x_n x$;

(b) if $x_n x = x'$ in A and x' is not an identity element then $p\rho_x = x_1 x_2 \ldots x_{n-1} x'$;

(c) if $x_n x = e_{i'}$ in A then $p\rho_x = x_1 x_2 \ldots x_{n-1}$ if $n > 1$ and $p\rho_x = e_{j'}$ if $n = 1$.

In all cases we obtain $p\rho_x \in S_{j'}$, so ρ_x is indeed a map from S_j to $S_{j'}$. For $x = e_i$ we also define ρ_x to be the identity map on $S_{i\sigma}$.

Consider now the map $\rho : A \to \mathcal{S}$ defined on vertices by $i \mapsto S_{i\sigma}$ and on edges by $x \mapsto \rho_x$. This is clearly a graph-map and it is a routine matter to check that $\rho_{xy} = \rho_x \rho_y$ whenever the product xy is defined in A. Thus ρ is a *representation* of A, i.e. a category-map from A to $\mathcal{S}$. Since ρ identifies any two vertices of A which are identified by σ it induces category-maps $\rho' : \vec{P}(A^\sigma) \to \mathcal{S}$ and $\rho'' : U_\sigma(A) \to \mathcal{S}$ such that the diagram

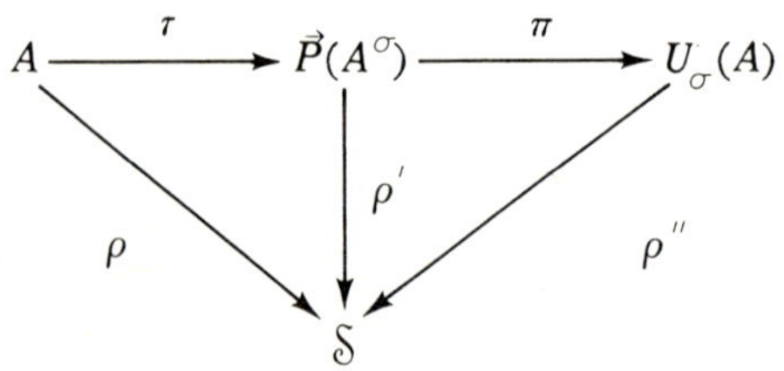

commutes, where the horizontal arrows are the canonical maps. (The fact that $\mathcal{S}$ is large is obviously irrelevant here since it can be replaced by a small subcategory). The reason for the commutativity relation $\pi\rho'' = \rho'$ is that $\tau\pi\rho'' = \rho = \tau\rho'$ and $A\tau$ generates $\vec{P}(A^\sigma)$.

To prove the theorem we have to show that distinct σ-reduced paths have distinct images in $U_\sigma(A)$ under the map π, so it is enough to show that they have distinct images in $\mathcal{S}$ under ρ'. By definition ρ' sends the edge x of A (or strictly, its image in A^σ) to the map ρ_x. Since ρ' is a category map it sends any path $p = x_1 x_2 \dots x_n$ in A^σ to the map $\rho_p = \rho_{x_1} \rho_{x_2} \dots \rho_{x_n}$ and we want to show that the maps ρ_p, for p σ-reduced, are all distinct. But if $p = x_1 x_2 \dots x_n$ $(n \geqslant 1)$ is a σ-reduced path from j to j' then the effect of the map ρ_p on the element e_j of S_j is to send it to $e_j \rho_{x_1} \rho_{x_2} \dots \rho_{x_n} = x_1 \rho_{x_2} \rho_{x_3} \dots \rho_{x_n} = \dots = x_1 x_2 \dots x_n = p$. If p is the empty path e_j, then ρ_p is the identity map on S_j, and again ρ_p sends e_j to p. Hence, if the σ-reduced paths p, q are such that $\rho_p = \rho_q : S_j \to S_{j'}$, then p, q have the same source j, and $p = e_j \rho_p = e_j \rho_q = q$. This proves the theorem. ■

Note. The theorem and its proof are valid for groupoids since the construction of $U_\sigma(A)$ in this case is exactly the same as for categories; the inverses take care of themselves. The theorem solves the word problem for $U_\sigma(A)$ (modulo that for A) since, if we want to know whether two paths in A^σ represent the same element of $U_\sigma(A)$ we need only apply successive elementary reductions to them until we arrive at σ-reduced paths, and check to see whether the reduced forms are the same or not. This can be done if we can tell (i) when products are defined in A and (ii) when two products in A give the same element of A. (Of course σ must be given in such a way that we can tell whether or not two vertices of A have the same image).

COROLLARY 1. *If two distinct edges of A have the same image in $U_\sigma(A)$ then they are identity elements at vertices i, i' such that $i\sigma = i'\sigma$.* ■

COROLLARY 2. *If B is a subcategory of A, and σ' is the restriction of σ to the vertices of B, then $U_{\sigma'}(B)$ can be identified with a subcategory of $U_\sigma(A)$.*

Proof. $B^{\sigma'}$ can be identified with a subgraph of A^σ, and then the σ'-reduced paths in $B^{\sigma'}$ are a subset of the σ-reduced paths in A^σ. ■

COROLLARY 3. ($\equiv$ *Proposition 9*). *If G is the free groupoid on a graph X then (i) $V(G) = V(X)$, (ii) X is embedded in G, (iii) every non-identity of G is uniquely expressible in the form $x_1^{\epsilon_1} x_2^{\epsilon_2} \ldots x_n^{\epsilon_n} (n \geqslant 1)$, where x_ν are edges of X, the ϵ_ν are ± 1, and no adjacent pairs xx^{-1} or $x^{-1}x$ occur, and (iv) the identities of G are not so expressible. Conversely, if G satisfies (i) (iii) and (iv) for a subgraph X, then X generates G freely.*

Proof. The free groupoid on X can be constructed as $U_\sigma(G_0)$, where G_0 is the disjoint union of simplicial groupoids G^x of type Δ^1, one for each edge x of X, and where σ is determined by the incidence maps of X. (See p.67, Example 2). We denote the non-identity edges of G^x by x and x^{-1}. Then the only products of non-identity edges which are defined in G_0 are the products xx^{-1} and $x^{-1}x$, so the σ-reduced paths in G_0 are precisely the empty paths and the paths $x_1^{\epsilon_1} x_2^{\epsilon_2} \ldots x_n^{\epsilon_n}$ described above. The corollary follows now from the theorem. ■

COROLLARY 4. *(The word problem for free products of groups and semigroups-with-1). Let $A^\lambda (\lambda \in \Lambda)$ and B be semigroups-with-1, and let $\theta^\lambda : A^\lambda \to B$ be homomorphisms (i.e. morphisms of $\mathcal{C}_1$). Then*

B is the free product of the A^λ relative to the maps θ^λ (i.e. their coproduct in $\mathcal{C}_1$) if and only if for every element $b \in B$ there is a unique sequence $\lambda_1, \lambda_2, \ldots, \lambda_n$ $(n \geqslant 0)$ in Λ, with $\lambda_\nu \neq \lambda_{\nu+1}$ $(i \leqslant \nu \leqslant n\text{-}1)$, and unique elements $a_\nu \neq 1$ in A^{λ_ν} $(1 \leqslant \nu \leqslant n)$ such that $b = b_1 b_2 \ldots b_n$, where $b_\nu = a_\nu \theta^{\lambda_\nu}$. The empty product stands for 1. In particular, the θ^λ are all injections and for $\lambda \neq \mu$ the images of A^λ and A^μ in B have only the identity in common. The same is true for free products of groups (coproducts in $\mathcal{G}_1$).

Proof. Let A be the disjoint union of the A^λ (their coproduct in $\mathcal{C}$ or $\mathcal{G}$, as the case may be). Then their coproduct in $\mathcal{C}_1$ or $\mathcal{G}_1$ is $U_\sigma(A)$, where σ maps the vertices of A to a single vertex. Assuming, as we may, that the A^λ are disjoint, the product aa' of edges of A is defined if and only if a and a' lie in the same A^λ. Consequently the σ-reduced words are precisely the empty path and the paths $a_1 a_2 \ldots a_n$ $(n \geqslant 1)$ satisfying the stated conditions.

Exercises

1. Show that $U_\sigma(A)$ is a groupoid if and only if A is a groupoid.

2. Show that a groupoid-map $\theta : A \to B$ is universal if and only if (i) $A\theta$ and the identities of B generate B and (ii) whenever $(a_1\theta)(a_2\theta) \ldots (a_n\theta) = 1$ $(n \geqslant 1)$ in B there exist $1 \leqslant r \leqslant s \leqslant n$ such that $a_r a_{r+1} \ldots a_s = 1$ in A. Here "$x_1 x_2 \ldots x_k = 1$" is short for "$x_1 x_2 \ldots x_k$ is defined and is an identity element".

CHAPTER 11

Free products of categories and groupoids

Coproducts of categories and groupoids are disjoint unions and might be called "absolute free products". The usual free product of groups is obtained from the absolute free product by imposing extra relations which equate all the identity elements and so might be called "free products with identities amalgamated". We shall use free products intermediate between these extremes, which include as a special case coproducts in the Categories $\mathcal{C}_I$ and $\mathcal{G}_I$. The solution of the word problem for these products is again contained in Theorem 4.†

Let $\theta^\lambda : A^\lambda \to B$ $(\lambda \in \Lambda)$ be category-maps with vertex maps $\sigma^\lambda : I^\lambda \to J$. We call B the *free product of the categories* A^λ (with respect to the maps θ^λ) if the following universal property holds: for any family of category-maps $\phi^\lambda : A^\lambda \to C$ whose vertex maps are of the form $\sigma^\lambda \tau : I^\lambda \to K$ (where $\tau : J \to K$ is independent of λ), there is a unique category-map $\phi : B \to C$ such that $\phi^\lambda = \theta^\lambda \phi$ for all $\lambda \in \Lambda$. Suppose that this is true, and write $A = \coprod A^\lambda$, $I = \coprod I^\lambda$. Then the maps θ^λ induce a map $\theta : A \to B$ with vertex map $\sigma : I \to J$ induced by the σ^λ. We claim that σ is surjective and that B is canonically isomorphic with $U_\sigma(A)$. For let $J_0 = I\sigma$, let $\sigma_0 : I \to J_0$ be the map induced by σ and let $\alpha : A \to U_{\sigma_0}(A)$ be the canonical map. If the A^λ are all empty it is easy to see that B is empty, and our assertion is valid. In all other cases J_0 is non-empty and there is at least

†See p.73. A list of theorems with page references is given in the index.

one map $\tau : J \to J_0$ such that $\sigma\tau = \sigma_0$. Hence, by the definition of free product, there is a unique category-map $\alpha^* : B \to U_{\sigma_0}(A)$ (induced by the restrictions of α to the A^λ) such that $\theta\alpha^* = \alpha$. Also there is a unique map $\mu : J_0 \to J$ such that $\sigma_0\mu = \sigma$, so θ induces a unique category-map $\theta^* : U_{\sigma_0}(A) \to B$ (with vertex map μ) such that $\alpha\theta^* = \theta$. It follows easily that α^* and θ^* are inverse isomorphisms. This shows that B is determined up to isomorphism by the A^λ and the σ^λ, and we write $B = *_{\lambda \in \Lambda}(A^\lambda; \sigma^\lambda)$, which will often be abbreviated to $B = *A^\lambda$. If the A^λ are subcategories of B and the θ^λ are the inclusion maps, we say that B is the *free product of subcategories* A^λ. Note that if the A^λ are all *groupoids* and $B = *A^\lambda$, then $B \cong U_{\sigma_0}(\coprod A^\lambda)$ is also a groupoid, and the universal property for free products holds in $\mathcal{G}$ as well as in $\mathcal{C}$. There is therefore no need to consider free products of groupoids separately except for special purposes.

If the A^λ and the $\sigma^\lambda : I^\lambda \to J$ are given, and if $J = \bigcup I^\lambda \sigma^\lambda$, then it is clear that $U_\sigma(\coprod A^\lambda)$ (where $\sigma : \coprod I^\lambda \to J$ is induced by the σ^λ) is in fact the free product of the A^λ with respect to the obvious maps. Corollary 2 of Theorem 4 shows that $U_{\sigma^\lambda}(A^\lambda)$ is embedded in this product. The subcategory generated by the image of A^λ coincides with this subcategory $U_{\sigma^\lambda}(A^\lambda)$ if σ^λ is surjective; otherwise it differs from it only in the deletion of certain isolated identity elements. We sum up in the following proposition; the reader can check any missing details for himself.

PROPOSITION 21. *(i) Let A^λ $(\lambda \in \Lambda)$ be categories (groupoids) with vertex sets I^λ and let $\sigma^\lambda : I^\lambda \to J$ be maps such that $J = \bigcup I^\lambda \sigma^\lambda$. Then the free product $*(A^\lambda; \sigma^\lambda)$ exists. (ii) Let $\theta^\lambda : A^\lambda \to B$ $(\lambda \in \Lambda)$ be arbitrary category-maps (groupoid-maps) and let B^λ be the subcategory (subgroupoid) of B generated by $A^\lambda\theta^\lambda$. Then $B = *A^\lambda$ with* *respect to the* θ^λ *if and only if the maps* θ^λ *are all universal and* B *is the free product of the subcategories (subgroupoids)* B^λ. ■

We may now restrict attention, for most purposes, to free products of subcategories or subgroupoids, and the next theorem gives various characterisations of this situation.

THEOREM 5. *Let* B^λ $(\lambda \in \Lambda)$ *be subcategories (subgroupoids) of the category (groupoid)* B, *and let* $V(B^\lambda) = J^\lambda$; $V(B) = J$. *Then the following are equivalent:*

(i) B *is the free product of the* B^λ;

(ii) if $\phi^\lambda : B^\lambda \to C$ *are morphisms which agree on common identity elements of the* B^λ, *there is a unique morphism* $\phi : B \to C$ *such that the restriction of* ϕ *to* B^λ *is* ϕ^λ *for all* $\lambda \in \Lambda$;

(iii) $\bigcup J^\lambda = J$; *for every non-identity element* b *of* B *there is a unique sequence* $\lambda_1, \lambda_2, \ldots, \lambda_n \in \Lambda$ *with* $\lambda_\nu \neq \lambda_{\nu+1}$ $(1 \leqslant \nu \leqslant n\text{-}1)$ *and unique non-identity elements* $b_\nu \in B^{\lambda_\nu}$ *such that* $b = b_1 b_2 \ldots b_n$; *and for identity elements* b *there is no such factorisation with* $n \geqslant 1$.

In the case of groupoids these are also equivalent to

(iv) the B^λ *generate* B, *and whenever* $b_1 b_2 \ldots b_n$ *is an identity of* B, *(where* $n \geqslant 1$, $b_\nu \in B^{\lambda_\nu}$ *and* $\lambda_\nu \neq \lambda_{\nu+1}$*), at least one of the* b_ν *is an identity.*

Proof. (ii) is a restatement of the definition of (i) in this special case. (i) is equivalent to the statement that the map $\coprod B^\lambda \to B$ induced by the inclusion maps is universal with surjective vertex map, and (iii) is a restatement of this in terms of the solution of the word problem given by Theorem 4. Also, (iii) implies (iv) trivially, for categories as well as for groupoids. Finally, in the groupoid case, suppose that (iv) is true. Then the first and last statements in (iii) are true. Also every non-identity b of B can be written as a product of edges of the B^λ and hence (by multiplying in the B^λ where possible and omitting identities) as a reduced word, i.e. as a product $b = b_1 b_2 \dots b_n$ of the stated type. If $b = c_1 c_2 \dots c_m$ is also a factorisation of this type, then $b_1 b_2 \dots b_n c_m^{-1} \dots c_2^{-1} c_1^{-1}$ is an identity of B, so by (iv), b_n and c_m^{-1} both lie in B^λ for some λ, and their product is an identity. Thus $c_m = b_n$, and induction shows that the two factorisations are the same. ■

PROPOSITION 22. *(The associative law for free products). Let $B^\lambda (\lambda \in \Lambda)$ be subcategories (subgroupoids) of the category (groupoid) B, and suppose that $\Lambda = \bigcup_{i \in I} \Lambda_i$, where the Λ_i are mutually disjoint. Let B^i be the subcategory (subgroupoid) of B generated by the B^λ for $\lambda \in \Lambda_i$. Then $B = *_{\lambda \in \Lambda} B^\lambda$ if and only if (i) $B^i = *_{\lambda \in \Lambda_i} B^\lambda$ for all $i \in I$ and (ii) $B = *_{i \in I} B^i$.*

Proof. Check the universal properties as stated in Theorem 5 (ii). ■

Similar results hold for arbitrary free products but are more troublesome to state.

It is clear that free products can be thought of as right limits. In the notation above, to say that B is the free product of the A^λ with respect to the θ^λ is to say that B is the right limit of the diagram whose objects are the A^λ and the trivial categories $T(I^\lambda)$, $T(J)$ and whose morphisms are the injections $T(I^\lambda) \to A^\lambda$ and the maps $T(\sigma^\lambda) : T(I^\lambda) \to T(J)$ (plus the assertion that $J = \bigcup I^\lambda \sigma^\lambda$). Or one can easily construct maps $\tau^\lambda : K \to I^\lambda$ whose right limit is J (with respect to the σ^λ) and then B is the right limit of the diagram with morphisms $T(\tau^\lambda) : T(K) \to A^\lambda$. Consequently the formation of free products commutes with other types of direct limit (see Proposition 18). We state some useful special cases and leave the reader to supply the details.

PROPOSITION 23. *(i) Let $\theta^\lambda : A^\lambda \to B$ be category-maps with vertex maps $\sigma^\lambda : I^\lambda \to J$, and let $\tau : J \to K$ be any map. If $B = *A^\lambda$ with respect to the θ^λ then $U_\tau(B) = *U_{\sigma^\lambda \tau}(A^\lambda)$ with respect to the maps induced by the θ^λ and τ. The same is true for groupoids.*

(ii) Let θ^λ, σ^λ be as in (i) and suppose that $A^\lambda = \mathrm{cat}(X^\lambda; R^\lambda)$. Let X be the graph with vertex set $\bigcup I^\lambda \sigma^\lambda \subset J$, and edge set $\coprod E(X^\lambda)$, with incidences determined by the maps $X^\lambda \to A^\lambda \to B$. Let R be the union of the images of the R^λ in $\vec{P}(X) \times \vec{P}(X)$. Then $B = *A^\lambda$ if and only if $B = \mathrm{cat}(X; R)$. The same is true for groupoids with the obvious changes of wording. ■

Exercises

1. Show that any free groupoid with no trivial components is the free product of subgroupoids which are either infinite cyclic groups or simplicial groupoids of type Δ^1.

2. Show that if the groupoid A is the free product of subgroupoids A^λ then, for $\lambda \neq \mu$, $A^\lambda \cap A^\mu$ is trivial.

3. Define free products of graphs by appropriate universal properties and determine their structure. Show that if $X^\lambda \to X$ $(\lambda \in \Lambda)$ is a free product diagram of graphs then $\pi(X^\lambda) \to \pi(X)$ $(\lambda \in \Lambda)$ is a free product diagram of groupoids (cf. Proposition 23 (ii)).

4. Suppose that the connected groupoid G is the free product of subgroupoids A and B. Prove that if A is a full subgroupoid then B is unicursal.

5. Show that the forgetful functor $\mathcal{G} \to \mathcal{C}$ has both a left and a right adjoint, and therefore preserves both left and right limits.

CHAPTER 12

Quotient maps of groupoids

From now on these notes will be concerned almost exclusively with groupoids. It is helpful to analyse groupoid-maps by defining classes of well-behaved maps and using these to factorise arbitrary maps. We have already met two such classes: the universal groupoid-maps and the (deformation) retractions. In the case of group-maps these reduce to isomorphisms and identity-maps respectively. We now introduce the class of quotient maps, which includes all group homomorphisms, and we shall see that most of the standard properties of group homomorphisms carry over to this larger class of groupoid-maps.

A subgroupoid N of a groupoid A is a *normal subgroupoid* if (i) N contains all the identity elements of A and (ii) $x \in N_{ii}$, $a \in A_{ij}$ implies $a^{-1}xa \in N_{jj}$. For any groupoid-map $\theta : A \to B$ we define the *kernel* of θ (denoted by Ker θ) to be the set of all elements of A which map to identity elements of B. Clearly Ker θ is always a normal subgroupoid of A, and we now show that every normal subgroupoid is the kernel of a groupoid-map. The situation differs from that of groups in that there will usually be many quite different groupoid-maps with the same kernel (for example all universal groupoid-maps have trivial kernel). The particular map we construct for a given kernel is the "best" one: it is a factor of all maps with the given kernel.

Let N be any normal subgroupoid of the groupoid A. The components of N define a partition on $I = V(A)$, and we write $\bar{i}$ for the class containing i, and $\bar{I}$ for the set of classes. N also defines an equivalence relation on $E(A)$ as follows: $a \equiv b \pmod N$ if and only if $a = xby$ for some $x, y \in N$. (This is clearly an equivalence relation for any subgroupoid N containing all identities of A). Two equivalent edges of A must have their sources in the same component of N, and similarly for their targets, so each class $\bar{a}$ of edges can be assigned a unique source and target in $\bar{I}$. This assignment defines a graph A/N, and the map $\pi : a \mapsto \bar{a},\ i \mapsto \bar{i}$ is a surjective graph-map from A to A/N. We now define a partial multiplication on the edges of A/N as follows: the product $\bar{a}\bar{b}$ is defined if and only if there exist $a_1 \in \bar{a},\ b_1 \in \bar{b}$ such that $a_1 b_1$ is defined in A, and then $\bar{a}\bar{b} = \overline{a_1 b_1}$. To see that this multiplication is well-defined, suppose that also $a_2 \in \bar{a},\ b_2 \in \bar{b}$ and $a_2 b_2$ is defined in A. Then $a_2 = x a_1 y$, $b_2 = z b_1 t$ with $x, y, z, t \in N$, and $a_2 b_2 = x a_1 y z b_1 t$ in A. Since $a_1 b_1$ is defined in A, yz lies in a vertex group of N, so $u = b_1^{-1} y z b_1$ is defined and lies in N. Hence $a_2 b_2 = x a_1 b_1 u t \equiv a_1 b_1 \pmod N$. (This argument, of course, depends on the normality of N). Now for $\bar{a} : \bar{i} \to \bar{j}$ and $\bar{b} : \bar{k} \to \bar{l}$ in A/N, the product $\bar{a}\bar{b}$ is defined if and only if axb is defined for some x in N, i.e. if and only if $\bar{j} = \bar{k}$, and then we have $\bar{a}\bar{b} = \overline{axb} : \bar{i} \to \bar{l}$. Moreover, if $(\bar{a}\bar{b})\bar{c}$ is defined, then $axbyc$ is defined for some $x, y \in N$, and the associative law in A/N follows immediately. Thus A/N is now a category whose identities are the components of N, and $\pi : A \to A/N$ is a category-map. It follows that A/N is a groupoid since A is a groupoid and π is surjective. We call $\pi : A \to A/N$ a *quotient map of groupoids*. Note that A/A is not in general the one-element groupoid; it is the trivial groupoid with one vertex for each component of A. Note also that if B is a subgroupoid of A containing N then B/N is a subgroupoid of A/N.

PROPOSITION 24. *Let $\pi : A \to A/N$ be a quotient map and let $\theta : A \to B$ be any groupoid-map with kernel $M \supset N$. Then there is a unique groupoid-map $\theta^* : A/N \to B$ such that $\theta = \pi\theta^*$. The kernel of θ^* is M/N. In particular, if* $\operatorname{Ker}\theta = N$ *then θ^* has trivial kernel.*

Proof. $a_1\pi = a_2\pi \Rightarrow a_1 \equiv a_2 (\mathrm{mod}\ N) \Rightarrow a_1 = x a_2 y\ (x, y \in N)$

$$\Rightarrow a_1\theta = (x\theta)(a_2\theta)(y\theta) = a_2\theta.$$

Thus there is a unique map $\theta^* : A/N \to B$ with $\theta = \pi\theta^*$, and it is easy to see that θ^* is a groupoid-map with the stated kernel. ∎

This universal property of A/N can be described in terms of push-outs as follows. For any groupoid G let $C(G)$ denote the set of components of G, and $TC(G)$ the trivial groupoid on $C(G)$. Then the canonical map $G \to TC(G)$ is universal amongst maps from G to trivial groupoids (in fact (C, T) is an adjoint pair of functors). Proposition 24 says that for a normal subgroupoid N of A the diagram

$$\begin{array}{ccc} N & \longrightarrow & A \\ \downarrow & & \downarrow \\ TC(N) & \longrightarrow & A/N \end{array}$$

is a pushout square. We shall say that any groupoid-map $\theta : A \to B$ is a *quotient map* if

$$\begin{array}{ccc} \operatorname{Ker}\theta & \longrightarrow & A \\ \downarrow & & \downarrow \theta \\ TC(\operatorname{Ker}\theta) & \longrightarrow & B \end{array}$$

is a pushout square. This is equivalent to the statement that the induced map $\theta^* : A_{/\operatorname{Ker}\theta} \to B$ is an isomorphism.

Any groupoid-map $\theta : A \to B$ induces a vertex map $V(\theta) : V(A) \to V(B)$, say $\sigma : I \to J$, and a family of edge maps $\theta_{ij} : A_{ij} \to B_{i\sigma, j\sigma}$ $(i,j \in I)$. It is convenient to classify groupoid-maps by properties of these induced maps. We shall say that θ is *vertex-surjective* if σ is a surjection, that θ is *piece-wise surjective* if each θ_{ij} is surjective, and that θ is *group-surjective* if each group homomorphism θ_{ii} is a surjection. The terms *vertex-injective, vertex-bijective,* etc. are similarly defined. Each of these classes of groupoid-maps is closed under composition and contains all isomorphisms. There are some obvious relations between them:

(i) θ is piece-wise injective if and only if it is group-injective, and this is equivalent to saying that Ker θ is unicursal;

(ii) If A is connected, then $\theta : A \to B$ is piece-wise surjective if and only if it is group-surjective;

(iii) θ is an injection if and only if it is vertex-injective and piece-wise injective;

(iv) if θ is vertex-surjective and piece-wise surjective then it is a surjection.

Note that the converse of (iv) is false: the non-trivial map from Δ^1 to a cylic group of order 2 is a counter-example. In fact we have:

PROPOSITION 25. *The following are equivalent for a groupoid-map $\theta : A \to B$ with kernel N:*

(i) θ is a quotient map;

(ii) θ is vertex surjective and piece-wise surjective;

(iii) θ is surjective, and any two vertices of A having the same image in B lie in the same component of N.

Proof. (i) $\Longrightarrow$ (ii) is clear from the construction of A/N.

(ii) $\Longrightarrow$ (iii). θ is certainly surjective. If i,j are vertices of A such that $i\theta = j\theta = k$, say, then $e_k \in B_{i\theta, j\theta}$ has a counter-image

y in A_{ij}, and $y \in N_{ij}$ since its image is an identity.

(iii) $\Longrightarrow$ (i). By Proposition 24, θ induces a unique groupoid-map $\theta^* : A/N \to B$, and θ^* has trivial kernel, i.e. is piece-wise injective. By (iii) the vertex map of θ^* is injective, since the vertices of A/N correspond to components of N. Thus θ^* is an injection. But θ is surjective, therefore θ^* is surjective, hence an isomorphism. ■

COROLLARY. *If $\theta : A \to B$ and $\phi : B \to C$ are quotient maps then so is $\theta\phi : A \to C$ (by criterion (ii)).* ■

This corollary can be stated in the more familiar form: if M, N are normal subgroupoids of A with $M \supset N$, then $(A/N)/(M/N) \cong A/M$. The other isomorphism theorems of group theory fail in general, but a special case is worth noting. Suppose that the normal subgroupoid N of A is totally disconnected (that is, its components are groups). Then the construction of A/N is simplified: $a \equiv b \pmod{N}$ if and only if a $a = bx$ for some x in N, and the quotient map $\pi : A \to A/N$ is vertex-bijective.

PROPOSITION 26. *Let $\theta : A \to B$ be a groupoid-map with kernel N. If θ is vertex-injective, then N is totally disconnected and $\theta^* : A/N \to B$ is an injection. Hence $A\theta$ is a subgroupoid of B isomorphic with A/N. In particular this is true for all morphisms in $\mathcal{G}_I$.*

Proof. N is obviously totally disconnected. θ^* is vertex-injective and has trivial kernel N/N, so is an injection. ■ (See Exercise 2 below for further results in this direction).

COROLLARY. *Every groupoid-map θ can be written $\theta = \theta_1 \theta_2 \theta_3$, where θ_1 is universal, θ_2 is a quotient-map and θ_3 is an injection. θ_1 may be chosen so that θ_2 and θ_3 are vertex-bijective.*

Proof. Take θ_1 the universal map induced by $V(\theta)$ and apply Proposition 26. ∎

Since quotient groupoids can be defined as pushouts, i.e. as right limits of suitable diagrams we can again formulate special cases of Proposition 18 which say that, in an appropriate sense, quotients commute with universals and with free products. For these we need the obvious notion of the normal subgroupoid of A generated by a subgraph X of A; it is the intersection of all normal subgroupoids containing X. (See Exercise 4 below for a description of its elements).

PROPOSITION 27. *(i) Let* $\mathbf{A} = \{A^\lambda\}$ *be a diagram in* $\mathfrak{G}$ *with right limit* L *and canonical maps* $\theta^\lambda : A^\lambda \to L$. *For each* λ, *let* N^λ *be a normal subgroupoid of* A^λ *such that every morphism* $A^\lambda \to A^\mu$ *in* $\mathbf{A}$ *maps* N^λ *into* N^μ. *Then the groupoids* A^λ/N^λ *are the objects of a diagram* $\mathbf{A}/\mathbf{N}$ *with morphisms induced by those of* $\mathbf{A}$. *If* M *is the normal subgroupoid of* L *generated by all the* $N^\lambda\theta^\lambda$, *then* L/M *is the right limit of* $\mathbf{A}/\mathbf{N}$ *with respect to the induced maps* $\overline{\theta}^\lambda : A^\lambda/N^\lambda \to L/M$.

(ii) Let $\theta : A \to B$ *be a universal groupoid-map, and let* $\alpha : A \to \overline{A}$ *be a quotient map with kernel* N. *Let* M *be the normal subgroupoid of* B *generated by* $N\theta$, *and let* $\beta : B \to \overline{B}$ *be a quotient map with kernel* M. *Then the induced map* $\overline{\theta} : \overline{A} \to \overline{B}$ *is universal.*

(iii) Let B *be the free product of groupoids* A^λ *with respect to maps* $\theta^\lambda : A^\lambda \to B$, *and let* $\alpha^\lambda : A^\lambda \to \overline{A}^\lambda$ *be quotient maps with kernels* N^λ. *Let* M *be the normal subgroupoid of* B *generated by all the* $N^\lambda\theta^\lambda$, *and let* $\beta : B \to \overline{B}$ *be a quotient map with kernel* M. *Then* $\overline{B}$ *is the free product of the* $\overline{A}^\lambda$ *with respect to the induced maps* $\overline{\theta}^\lambda : \overline{A}^\lambda \to \overline{B}$.

Proof. We leave this as an exercise. Either apply Proposition 18 to suitable diagrams or proceed directly from the particular universal properties involved. ∎

COROLLARY. *Let A be the free product of subgroupoids A^λ, let N^λ be a normal subgroupoid of A^λ for each λ, and let N be the normal subgroupoid of A generated by all the N^λ. If $C = A/N$ and $\pi : A \to C$ is the quotient map, then $C = *C^\lambda$, where C^λ is the subgroupoid generated by $A^\lambda\pi$, and the induced maps $A^\lambda/N^\lambda \to C^\lambda$ are universal. Moreover, $N \cap A^\lambda = N^\lambda$ for all λ.*

Proof. By part (iii) of the proposition, with θ^λ the inclusion map $A^\lambda \to A$, C is the free product of the groupoids A^λ/N^λ with respect to the induced maps $A^\lambda/N^\lambda \to C$. By Proposition 21(ii), this implies that $C = *C^\lambda$ and that $A^\lambda/N^\lambda \to C^\lambda$ is universal. The kernel of $A^\lambda/N^\lambda \to C^\lambda$ is $(N \cap A^\lambda)/N^\lambda$, and since universal morphisms have trivial kernel (Theorem 4, Cor. 1, p.76) it follows that $N \cap A^\lambda = N^\lambda$ for all λ. ∎

We now apply these ideas to obtain a detailed analysis of retractions. We recall that a retraction is a groupoid-map $\rho : A \to B$, where B is a subgroupoid of A with inclusion map $\mu : B \to A$, such that $\mu\rho = 1_B$ and $\rho\mu \simeq 1_A$ (see p.47).

PROPOSITION 28. *Let B be a subgroupoid of the groupoid A with inclusion map $\mu : B \to A$. Let $\rho : A \to B$ be a groupoid-map with kernel N. Then the following are equivalent:*

(i) ρ is a retraction;

(ii) ρ is piecewise bijective and $\mu\rho = 1_B$;

(iii) ρ is a quotient map, N is unicursal, and $\mu\rho = 1_B$.

Proof. (i) $\Longrightarrow$ (ii). Let $\sigma : I \to K$ be the vertex map of ρ. Since $\rho\mu \simeq 1_A$, there exist elements $r_i : i \to i\sigma$ of A (for all $i \in I$) such that, for any $x \in A_{ij}$, $x\rho = r_i^{-1} x r_j$. It follows that ρ maps A_{ij} bijectively to $A_{i\sigma, j\sigma}$, the inverse map being given by $y \mapsto r_i y r_j^{-1}$ for $y \in A_{i\sigma, j\sigma}$. Hence B is a full subgroupoid and ρ is piecewise bijective.

(ii) $\Longrightarrow$ (iii). Since $\mu\rho = 1_B$, ρ is surjective. Since ρ is also piecewise surjective, it is a quotient map (Proposition 25). Since ρ is piecewise injective, N is unicursal.

(iii) $\Longrightarrow$ (i). Let $x \in A_{ij}$, and consider $x\rho \in B_{i\sigma, j\sigma}$. Since $\mu\rho = 1_B$, x and $x\rho$ have the same image under ρ. But ρ is a quotient map, so this implies $x\rho \equiv x \pmod{N}$. Hence $x\rho = r_i^{-1} x r_j$ for some $r_i : i \to i\sigma$ and $r_j : j \to j\sigma$ in N. Since N is unicursal, r_i is uniquely determined (independently of x) for each $i \in I$, and the case $x = e_i$ shows that r_i exists for each i. The family $\{r_i\}_{i \in I}$ now gives the required homotopy $\rho\mu \simeq 1_A$. ∎

THEOREM 6. *(i) Let $\rho : A \to B$ be a retraction with kernel N. Then N is unicursal, B is a full subgroupoid containing exactly one vertex of each component of N, and $A = B * N$ (free product of subgroupoids). Moreover, if B' is any subgroupoid of A such that $A = B' * N$ then the map $B' \to B$ induced by ρ is universal.*

(ii) Let N be any unicursal subgroupoid of A containing all vertices of A. (N is then normal). Let B be any full subgroupoid of A containing exactly one vertex of each component of N. Then $A = B * N$ and there is a unique retraction $A \to B$ with kernel N.

Proof. We prove (ii) first. Suppose that N and B are as stated there. For each vertex i of A, let i' denote the unique vertex of B lying in the same component of N as i does, and let r_i be the unique element of $N_{ii'}$. For $x \in A_{ij}$ we have $x' = r_i^{-1} x r_j \in A_{i'j'} = B_{i'j'}$, and the map $\rho : x \mapsto x'$ is clearly a retraction from A to B since if $x \in B_{kl}$

then $r_k = e_k$, $r_l = e_l$ and $x' = x$. To show that $A = B * N$ we use criterion (ii) of Theorem 5 (p.81). Suppose that $\phi : B \to C$ and $\psi : N \to C$ are groupoid maps which agree on the vertices of B. We look for a groupoid-map $\theta : A \to C$ whose restrictions to B and N are ϕ and ψ respectively. For $x \in A_{ij}$, we have $x = r_i(x\rho)r_j^{-1}$, and since $x\rho \in B$ we have no choice but to define $x\theta = (r_i\psi)(x\rho\phi)(r_j\psi)^{-1}$. It is easy to check that this gives a groupoid-map $\theta : A \to C$ with the required properties, and it is clearly unique. Thus $A = B * N$. The uniqueness of ρ follows from this since any two retractions $A \to B$ with kernel N induce the same map $B \to B$ (namely 1_B) and the same map $N \to B$ (each component of N maps to the identity element of B lying in it).

To prove (i), let $\rho : A \to B$ be any retraction with kernel N. We already know that B is then a full subgroupoid of A, and N is unicursal (see the proof of Proposition 28). Also, since ρ is a quotient map and induces the identity map on B, each congruence class (mod N) contains exactly one edge of B, so each component of N contains exactly one identity of B. It follows from (ii) that $A = B * N$. Suppose, finally, that we also have $A = B' * N$. The trivial normal subgroupoid N' of B' and the normal subgroupoid N of N together generate the normal subgroupoid N of A. Hence, by the corollary to Proposition 27, the map $\rho : A \to B$ (which is a quotient map with kernel N) induces a decomposition $B = B_1 * B_2$, where $B_1 = \text{gpd}\{B'\rho\}$ and $B_2 = \text{gpd}\{N\rho\}$, and the induced map $B' = B'/N' \to B_1$ is universal. But B_2 contains only the identity elements of B, so clearly the inclusion map $B_1 \to B$ is universal, and it follows (Proposition 19') that the composite map $B' \to B_1 \to B$ is universal. ■

COROLLARY 1. *Every groupoid* A *has a free decomposition* $A = B * N$, *where* B *is a totally disconnected subgroupoid consisting of one (arbitrary) vertex group from each component of* A, *and* N *is a unicursal subgroupoid generated by an arbitrary maximal circuit-free subgraph of* A. *In particular, if* A *is connected then* $A = B * N$, *where* B *is an arbitrary vertex group of* A *and* N *is a simplicial groupoid generated by an arbitrary maximal tree in* A.

Proof. Choose a maximal circuit-free graph X in A. By Theorem 1, such a graph exists and spans A. By Proposition 10, Corollary 2, the subgroupoid N generated by X is unicursal, and it spans A. If A is connected then X is a tree and N is simplicial. Now any subgroupoid B consisting of vertex groups, one from each component of A, is a full subgroupoid and has one vertex in each component of N. The result now follows from part (ii) of the theorem. (See also Theorem 2, p.47). ■

COROLLARY 2. *If* $G = \text{gpd}(X; R)$ *is connected and* T *is any maximal tree in* G, *then the vertex group* G_{ii} *of* G *at the vertex* i *has a presentation* $G_{ii} = \text{gpd}(X; R \cup T^*)$, *where* T^* *is the set of all pairs* (t, e_i) *with* t *an edge of* T *written as a word in* X.

Proof. Let N be the normal subgroupoid generated by T. By part (ii) of the theorem, $G = G_{ii} * N$ and $G_{ii} \cong G/N$. The result follows from the universal properties of quotients and presentations. ■

Exercises

1. Let $\theta : A \to B$ be a quotient map and C a subgroupoid of A. Show that the induced map $C \to C\theta$ is a quotient map if and only if each component of $N = \text{Ker}\ \theta$ contains at most one component of $C \cap N$.

2. Show that in the Category $\mathcal{G}_I$ all the isomorphism theorems of group theory are valid (including the Zassenhaus lemma and hence the Schreier refinement theorem). All normal subgroupoids appearing must be totally disconnected.

3. Show that any groupoid-map θ can be written $\theta = \theta_1 \theta_2 \theta_3$, where θ_1 is vertex-injective, θ_2 is a retraction and θ_3 has trivial kernel.

4. Let X be a subgraph of the groupoid A. Let H be the subgroupoid generated by X and let $\overline{H}$ be the subgroupoid generated by H and all conjugates $a^{-1}ha$ of elements h of H. Show that $\overline{H}$ is the normal subgroupoid of A generated by X. (N.B. This is not as obvious as it looks. The trouble is that only elements of vertex groups have conjugates, and it is certainly not enough to take the subgroupoid generated by X and its conjugates as in the group case).

5. Let α, β be groupoid-maps, and call β a push-out of α if there exists a push-out square

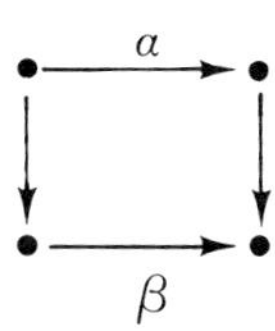

in $\mathcal{G}$. Show that push-outs of universal maps are universal and push-outs of quotient maps are quotient maps.

6. Let A be a free product of subgroupoids $A = B * C$, let N be the normal subgroupoid of A generated by C, and let $\pi : A \to \overline{A}$ be a quotient map with kernel N. Prove that (i) the induced map $B \to \overline{A}$ is universal, and (ii) there is a subgroupoid B' of A such that $A = B' * C$ and the induced map $B' \to \overline{A}$ is an *isomorphism.*

7. Prove that the vertex groups of a free groupoid are free groups.

CHAPTER 13

Covering maps

The covering spaces of a topological space T can be classified by their fundamental groupoids. The properties of the groupoids which make this possible are as follows. Let $\tilde{T} \to T$ be a covering map and $\gamma : \pi(\tilde{T}) \to \pi(T)$ the corresponding map of fundamental groupoids. Let i be any vertex of $\pi(T)$ (i.e. a point of T) and let $\tilde{i}$ be any vertex of $\pi(\tilde{T})$ with $\tilde{i}\gamma = i$. Then for each edge x of $\pi(T)$ with source i, there is exactly one edge $\tilde{x}$ of $\pi(\tilde{T})$ with source $\tilde{i}$ such that $\tilde{x}\gamma = x$. We take this property of the groupoid map γ as the definition of covering maps of groupoids, and extend it to include coverings of graphs and categories. These coverings have interesting algebraic properties.

Let A be a graph with vertex set I. The *star* of A at the vertex i is the set $\bigcup_{j \in I} A_{ij}$ of all edges of A with source i. We denote it by A_{i*} or $St_i(A)$. If $\theta : A \to B$ is a graph-map then θ induces star maps $\theta_{i*} : A_{i*} \to B_{k*}$ (where $k = i\theta$) for each $i \in I$. We say that θ is *star-injective* (*star-surjective, star-bijective*) if each of the maps θ_{i*} is injective (surjective, bijective). In this terminology a *covering map* of graphs is a star-bijective graph-map. Similarly, a covering map of categories or groupoids is a star-bijective category-map or groupoid-map. There is a dual notion of *co-covering* i.e. a morphism which maps all the co-stars $A_{*j} = \bigcup_{i \in I} A_{ij}$ bijectively.

For groupoids the two notions coincide since we have an involution $x \mapsto x^{-1}$ which interchanges stars and co-stars.

Notes: (i) We do not insist that coverings $\theta : A \to B$ should be surjective, nor that A and B should be connected, as is sometimes assumed for coverings of topological spaces.

(ii) Groupoid-maps which are star-surjective might reasonably be called fibrations; they imitate closely the basic properties of topological fibrations (see Exercises 2, 3, p.115). We shall not discuss them here apart from noting that fibrations can be characterised as quotient maps followed by covering maps. This makes them superfluous for our present purposes. (See also Brown [7]).

Examples. 1. Let T be a trivial category and A any category. Then the projection map $A \times T \to A$ is a covering (and a co-covering).

2. The inclusion map $C \to A$, where C is a component of the groupoid A, is a covering.

3. The non-trivial groupoid-map from the simplicial groupoid Δ^1 to the cyclic group of order 2 is a covering.

4. Let G be a group, and let $\hat{G}$ be the simplicial groupoid with vertex set G and edges (g, h) $(g, h \in G)$, with multiplication $(g, h)(h, k) = (g, k)$. Then the map $\gamma : \hat{G} \to G$ given by $(g, h) \mapsto g^{-1}h$ is a covering map of groupoids. (Example 3 is a special case). $\hat{G}$ is the *universal cover* of G.

PROPOSITION 29. *Pull-backs of coverings are coverings; that is if*

$$\begin{array}{ccc} \tilde{A} & \xrightarrow{\tilde{\theta}} & \tilde{B} \\ {\scriptstyle \alpha}\downarrow & & \downarrow{\scriptstyle \beta} \\ A & \xrightarrow[\theta]{} & B \end{array}$$

is a pull-back square in $\mathfrak{D}$, $\mathfrak{C}$ *or* $\mathfrak{G}$, *and if* β *is a covering morphism, then so is* α.

Proof. The construction of pull-backs is the same in all three Categories. The vertices of $\tilde{A}$ are pairs $(i, \tilde{j})(i \in V(A),\ \tilde{j} \in V(\tilde{B}))$ such that $i\theta = \tilde{j}\beta$. The edges of $\tilde{A}$ with source $(i, \tilde{j})$ are pairs $(a, \tilde{b})$ $(a \in A_{i*},\ \tilde{b} \in \tilde{B}_{\tilde{j}*})$ such that $a\theta = \tilde{b}\beta$. The vertex $(i, \tilde{j})$ of $\tilde{A}$ lies over the vertex i of A, and for any $a \in A_{i*}$, $A\,\theta \in B_{j*}$, where $j = i\theta = \tilde{j}\beta$. Since β is a covering, there is a unique $\tilde{b}$ in the star of $\tilde{B}$ at $\tilde{j}$ such that $\tilde{b}\beta = a\theta$, and $(a, \tilde{b})$ is then the unique element lying over a in the star of $\tilde{A}$ at $(i, \tilde{j})$. ■

Given β and θ as above, we call α the *induced covering* of A.

There is a close connection between covering maps $\tilde{A} \to A$ and representations of A. We define a *representation* of a graph A to be a graph-map $\phi : A \to \mathfrak{S}$, i.e. an A-diagram in the Category of sets. A *representation* of a category or groupoid A is a *functor* $\phi : A \to \mathfrak{S}$. The representations of a fixed A are the objects of a Category $\mathfrak{S}^A$ whose morphisms are the natural transformations (or morphisms of A-diagrams), and two representations are *equivalent* if they are isomorphic in this Category. When A is a group this definition coincides with the usual definition of equivalence for permutational representations of A. Similarly we can form a Category Cov(A) from the coverings of A. The objects of Cov(A) are all covering morphisms $\beta : B \to A$, and the morphisms from β to γ are all commutative diagrams

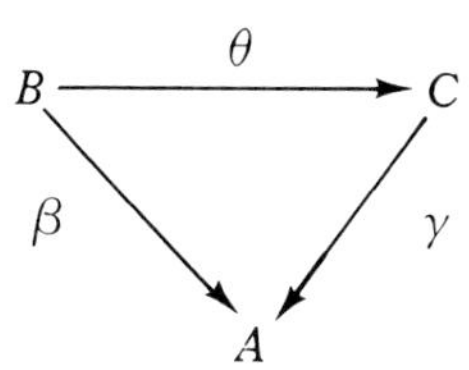

where θ is a morphism of the appropriate Category $\mathfrak{D}$, $\mathfrak{C}$ or $\mathfrak{G}$. Two coverings β, γ as above are *equivalent* if there is an isomorphism $\theta : B \to C$ with $\theta\gamma = \beta$, i.e. if β, γ are isomorphic objects of Cov(A). We shall show that the Categories $\mathfrak{S}^A$ and Cov(A) are equivalent in a very natural way.

Suppose that $\beta : B \to A$ is a covering map and let $I = V(A)$. For $i \in I$ write $B^i = i\beta^{-1} = \{v \in V(B) \mid v\beta = i\}$. If $a \in A_{ij}$ then (by definition of covering maps) there is, for each $v \in B^i$, a unique edge $\bar{a}$ with source v such that $\bar{a}\beta = a$. Denoting the target of $\bar{a}$ by w, we have $w\beta = j$, so the assignation $v \mapsto w$ defines a map $\beta^a : B^i \to B^j$, uniquely determined by $a \in A_{ij}$. The covering β therefore determines a graph-map $\beta^* : A \to \mathfrak{S}$ defined by $i \mapsto B^i$, $a \mapsto \beta^a$. Further, if β is a covering of categories or groupoids, it is clear that β^* is a category-map. Note that we may have $B^i = \emptyset$ for some i, in which case every $a \in A_{ij}$ is represented by the empty map $\emptyset \to B^j$.

If $\gamma : C \to A$ is another covering of A with corresponding representation $\gamma^* : i \to C^i$, $a \mapsto \gamma^a$, and if $\theta : B \to C$ is a morphism such that $\theta\gamma = \beta$, then θ induces maps $\theta^i : B^i \to C^i$. Also if $a \in A_{ij}$ and $\bar{a}$ is the edge of B lying over it with source v, then $\bar{a}\theta$ is the unique edge of C lying over a with source $v\theta$, so $\beta^a\theta^j = \theta^i\gamma^a$, and $\{\theta^i\}_{i \in I}$ is a natural transformation from β^* to γ^*. After a little more checking we see that we have a functor Cov$(A) \to \mathfrak{S}^A$ defined by $\beta \mapsto \beta^*$, $\theta \mapsto \{\theta^i\}_{i \in I}$.

Conversely, suppose we have a representation $\sigma : A \to \mathfrak{S}$ given by $i \mapsto S^i$, $a \to \sigma^a$, where $\sigma^a : S^i \to S^j$ if $a \in A_{ij}$. We define a covering $\sigma_* : S \to A$ as follows. The vertex set of S is $\coprod_{i \in I} S^i$, which we take to be the set of all pairs (i, v) with $v \in S^i$. The edges of S from (i, v) to (j, w) are all pairs (a, v) where $a \in A_{ij}$ and σ^a maps v to w. $\sigma_* : S \to A$ is defined by $(i, v) \mapsto i$, $(a, v) \mapsto a$, and the reader will easily verify that this is a covering morphism. If $\tau : A \to \mathfrak{S}$ is another representation and $\theta : \sigma \to \tau$ is a natural transformation given by maps $\theta^i : S^i \to T^i$ we obtain a morphism $\theta_* : S \to T$ by mapping the vertex (i, v) to $(i, v\theta^i)$ and the edge (a, v) $(a \in A_{ij})$ to the edge $(a, v\theta^i)$. This θ_* defines a morphism $\sigma_* \to \tau_*$ in Cov(A), and we now have a functor $\mathcal{S}^A \to \text{Cov}(A)$ given by $\sigma \mapsto \sigma_*$, $\{\theta^i\}_{i \in I} \mapsto \theta_*$.

If we start with a representation $\sigma : A \to \mathcal{S}$, pass to the covering σ_* and then to the induced representation $(\sigma_*)^* = \bar{\sigma}$, say, we find that the sets S^i are replaced by the disjoint sets $\bar{S}^i = \{(i, v) \mid v \in S^i\}$ and $\bar{\sigma}$ is equivalent to σ. It is obvious that the functor $\sigma \mapsto \bar{\sigma}$ from $\mathcal{S}^A$ to $\mathcal{S}^A$ is naturally isomorphic to the identity functor. Similarly, if we start with a covering $\beta : B \to A$ and form the new covering $(\beta^*)_*$ we find that the vertices of B are now provided with (superfluous) labels indicating their images in A, and the edges of B are replaced by an indication of their images in A and their sources in B. It is again clear that the functor $\beta \mapsto (\beta^*)_*$ is naturally isomorphic to the identity functor on Cov(B). This proves the equivalence of the two Categories, and the reader will have no difficulty in proving the last part of the following:

PROPOSITION 30. *Let A be a graph, a category or a groupoid. Then the Categories $\mathcal{S}^A$ and* Cov(A) *are equivalent. If $\sigma : A \to \mathcal{S}$ is a representation with corresponding cover $\sigma_* : S \to A$, and if $\theta : B \to A$ is any morphism (in $\mathfrak{D}$, $\mathfrak{C}$ or $\mathfrak{G}$) then the induced covering of B (the pull-back of σ_*) corresponds to the representation $\theta\sigma : B \to \mathcal{S}$ of B. (Hence $\mathcal{S}^A$ and* Cov(A), *thought of as contravariant Functors of A, are naturally equivalent).* ■

It is interesting to note that there is a prototype "Covering" which induces all coverings. Let $\tilde{\mathcal{S}}$ denote the Category of sets with base-point. Its objects are pairs (S, s), where S is a set and $s \in S$. Its morphisms from (S, s) to (T, t) are pairs (θ, s), where θ is a map from S to T such that $s\theta = t$. The forgetful functor $F : \tilde{\mathcal{S}} \to \mathcal{S}$ is a covering of $\mathcal{S}$; it corresponds to the identity representation of $\mathcal{S}$. If now $\sigma : A \to \mathcal{S}$ is a representation of A then the covering of A induced by F is (equivalent to) the covering σ_* defined above.

Coverings of groupoids have most of the formal properties of coverings of topological spaces, and we now give some examples. We can usually reduce to the case of *connected coverings,* i.e. coverings in which both base groupoid and covering groupoid are connected and non-empty. These correspond to representations which are *transitive* in an obvious sense.

PROPOSITION 31. *Let $\alpha : \tilde{A} \to A$ be a covering map of groupoids. Then*

(i) Ker *α is trivial, whence α is group-injective;*

(ii) each component of $\tilde{A}$ covers a component of A (but several components of $\tilde{A}$ may cover the same component of A);

(iii) if α is a connected covering then it is surjective;

(iv) if α is connected and maps some vertex group of $\tilde{A}$ surjectively to a vertex group of A then α is an isomorphism.

Proof. (i) If $x \in \tilde{A}_{ij}$ is in Ker α then $x\alpha$ is an identity of A and must be equal to $e_i\alpha$. Since α is star-injective this implies $x = e_i$. Group-maps with trivial kernel are injections.

(ii) Any groupoid-map $\alpha : \tilde{A} \to A$ sends each component $\tilde{C}$ of $\tilde{A}$ into a component C of A. Since each star of $\tilde{A}$ or A is contained in a component, the induced map $\tilde{C} \to C$ is still star-bijective.

(iii) If α is connected then $\tilde{A}$ is not empty, so at least one star of A is in the image of α. Since A is connected, it follows that all vertices of A are in $\tilde{A}\alpha$ and hence all stars of A are in $\tilde{A}\alpha$.

(iv) Let the vertex group $\tilde{A}_{00}$ be mapped surjectively (therefore isomorphically) to the vertex group A_{00} by α. Suppose that the vertices i, j of $\tilde{A}$ map to the same vertex k of A. Since $\tilde{A}$ is connected there are edges $\tilde{x} : 0 \to i$ and $\tilde{y} : 0 \to j$ in $\tilde{A}$. Their images x, y in A both lie in A_{0k}, so $z = xy^{-1} \in A_{00}$. By hypothesis, there is a $\tilde{z} \in \tilde{A}_{00}$ such that $\tilde{z}\alpha = z$. Consider the edge $\tilde{x}^{-1}\tilde{z}\tilde{y} : i \to j$ in $\tilde{A}$. Its image under α is $x^{-1}zy = e_k$. By (i) it must be an identity of $\tilde{A}$, so we have $i = j$. Thus α is vertex-injective and therefore injective (since it is star-injective). By (iii) α is also surjective, so it is an isomorphism. ∎

Note. If the covering $\alpha : \tilde{A} \to A$ is group-surjective (therefore group-bijective) then $\tilde{A}$ consists of a number (possibly 0) of disjoint isomorphic copies of the various components of A. Such coverings are called *trivial;* they correspond to *trivial representations* of A in which all edges are represented by identity maps (one identity map for each component of A). Of course they also correspond to representations equivalent to such trivial ones i.e. to representations which map A into a unicursal subgroupoid of $\mathcal{S}$.

PROPOSITION 32. *If $\alpha : \tilde{A} \to A$ is a covering map of groupoids then the fibres $a\alpha^{-1}$ as a runs through a component of A all have the same cardinal.*

Proof. In the corresponding representation of A each $a \in A_{ij}$ is represented by an invertible map $S_i \to S_j$, so the S_i all have the same cardinal for i in a fixed component. These are the fibres over the vertices. The fibre over $a \in A_{ij}$ has exactly one element for each vertex over i. ∎

If α is a connected covering and its fibres have finite cardinal n, we call α an *n-fold covering.*

PROPOSITION 33. *Let* $A \xrightarrow{\theta} B \xrightarrow{\phi} C$ *be morphisms in* $\mathfrak{D}$, $\mathfrak{C}$ *or* $\mathfrak{G}$. *If* θ *and* ϕ *are coverings, so is* $\theta\phi$. *If* ϕ *and* $\theta\phi$ *are coverings, so is* θ. *If* θ *and* $\theta\phi$ *are coverings and* θ *is surjective, then* ϕ *is a covering.*

Proof. This is an immediate consequence of the definition of coverings. ■

Next we turn to "homotopy lifting properties" of covering maps of groupoids. The analogue for groupoids of the closed unit interval of homotopy theory is the simplicial groupoid Δ^1 (the absolute free groupoid on one generator).

We shall denote the vertices of Δ^1 by 0, 1 and its edges by e_0, e_1, s, s^{-1}, where $s : 0 \to 1$. If G is any groupoid there are two canonical embeddings $\mu_0, \mu_1 : G \to G \times \Delta^1$ defined on vertices by $i \mapsto (i, t)$ and on edges by $g \mapsto (g, e_t)$ $(t = 0,1)$. These two maps are homotopic (naturally equivalent) under the natural transformation $\sigma : \mu_0 \to \mu_1$ given by the edges $s_i = (e_i, s)$ of $G \times \Delta^1$ for $i \in I = V(G)$; for if $g \in G_{ij}$ then $s_i^{-1}(g\mu_0)s_j = (e_i, s^{-1})(g, e_0)(e_j, s) = g\mu_1$. Consequently any groupoid-map $\theta : G \times \Delta^1 \to A$ defines a homotopy $a : \theta_0 \to \theta_1$ where $\theta_t = \mu_t\theta\,(t = 0,1)$, and a is given by the edges $s_i\theta$ of A. Conversely, if $\theta_0, \theta_1 : G \to A$ are groupoid-maps, and $a : \theta_0 \to \theta_1$ is a homotopy given by edges $a_i : i\theta_0 \to i\theta_1$ of A, then there is a unique groupoid-map $\theta : G \times \Delta^1 \to A$ such that $\theta_t = \mu_t\theta$ $(t = 0,1)$ and $s_1\theta = a_i$ $(i \in I)$. This can be checked directly, but the following argument is instructive. To simplify notation we shall denote a trivial groupoid by the same symbol as its vertex set, and in the case of a trivial group, by the same symbol as its vertex. $G \times \Delta^1$ then has subgroupoids $G \times 0$, $G \times 1$ which are the images of G under μ_0, μ_1, and a subgroupoid $I \times \Delta^1$ which is unicursal. Now

$G \times 0$ is a full subgroupoid and meets each component of $I \times \Delta^1$ in one vertex. By Theorem 6 (p.92) we therefore have $G \times \Delta^1 = (G \times 0) * (I \times \Delta^1)$. Now $\theta_0 : G \to A$ induces a map $\theta_0^* : G \times 0 \to A$, and since $I \times \Delta^1$ is the absolute free groupoid generated by the edges $s_i = (i, s)$, any choice of edges a_i in A gives a groupoid-map $\alpha^* : I \times \Delta^1 \to A$ with $s_i \alpha^* = a_i$. In our case a_i has source $i\theta_0$, so θ_0^* and α^* agree on the vertices of $G \times 0$ and therefore induce a groupoid-map $\theta : G \times \Delta^1 \to A$ with the required properties. (The fact that $\mu_1 \theta = \theta_1$ follows from the equations $g\theta_1 = a_i^{-1}(g\theta_0)a_j$ for $g \in G_{ij}$ which assert that $\alpha : \theta_0 \to \theta_1$ is a homotopy). We are now justified in speaking of homotopies $G \times \Delta^1 \to A$.

Note. For category-maps $\theta_0, \theta_1 : C \to A$, natural transformations $\theta_0 \to \theta_1$ are given by category-maps $\theta : C \times \{1\} \to A$ where $\{1\}$ is the absolute free category on one generator. Natural equivalences are given by maps $\theta : C \times \Delta^1 \to A$, and a modified version of the above argument can still be used in this case.

We want to investigate the question when a groupoid-map $\theta : G \to A$ can be lifted to a covering of A. If $\alpha : \tilde{A} \to A$ is a covering map, we say that θ can be *lifted* to $\tilde{A}$ if there is a groupoid-map $\tilde{\theta} : G \to \tilde{A}$ with $\tilde{\theta}\alpha = \theta$, and we then call $\tilde{\theta}$ a *lifting* of θ.

PROPOSITION 34. *(Unique homotopy lifting). Let $\alpha : \tilde{A} \to A$ be a covering map of groupoids, let $\theta : G \times \Delta^1 \to A$ be a homotopy and let $\mu_0 : G \to G \times \Delta^1$ be the canonical map with image $G \times 0$. If the map $\theta_0 = \mu_0\theta : G \to A$ has a lifting $\tilde{\theta}_0 : G \to \tilde{A}$, then θ has a unique lifting $\tilde{\theta} : G \times \Delta^1 \to \tilde{A}$ which makes the following diagram commutative*

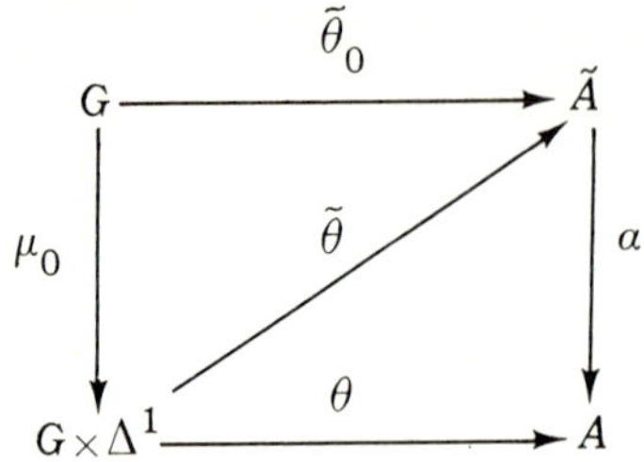

Proof. First note that when G is the trivial group, this assertion is precisely the definition of a covering map, since mapping Δ^1 into a groupoid is the same as choosing an edge. Similarly, if G is a trivial groupoid, $G \times \Delta^1$ is the disjoint union of copies of Δ^1 and the assertion follows immediately. In the general case, let $I = V(G)$ and apply these remarks to $I \times \Delta^1$. We then obtain a unique groupoid-map $\tilde{\phi} : I \times \Delta^1 \to A$ which makes the diagram

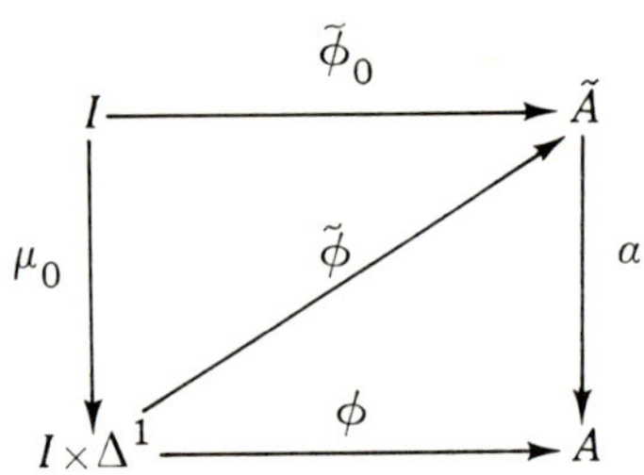

commutative, where $\tilde{\phi}_0$, ϕ are restrictions of $\tilde{\theta}_0$, θ. But, as shown above, $G \times \Delta^1$ is the free product of $G \times 0$ and $I \times \Delta^1$. If we think of $\tilde{\theta}_0$ as a map from $G \times 0$ to $\tilde{A}$, it agrees with $\tilde{\phi}$ on the vertex set $I \times 0$ of $G \times 0$; hence there is a unique groupoid-map $\tilde{\theta} : G \times \Delta \to A$ which restricts to $\tilde{\theta}_0$ and $\tilde{\phi}$ on $G \times 0$ and $I \times \Delta^1$. It is easy to see that θ has the required properties; its uniqueness follows from that of $\tilde{\phi}$. ■

To answer the general lifting problem for coverings we need a stronger result than this. Its proof runs along the same lines.

PROPOSITION 35. *Let $\alpha : \tilde{A} \to A$ be a covering of groupoids and let $\theta : G \to A$ be a groupoid-map. Let H be a retract of G (i.e. a full subgroupoid meeting each component of G), and let θ_0 be the restriction of θ to H. If θ_0 lifts to $\tilde{\theta}_0 : H \to \tilde{A}$ then θ has a unique lifting $\tilde{\theta} : G \to \tilde{A}$ which extends $\tilde{\theta}_0$.*

Proof. By Theorem 6 (p.92), $G = H * N$, where N is a unicursal subgroupoid of G, and H contains exactly one vertex of each component of N. Consider first the case when H is a trivial group. Then $G = N$ is simplicial and H lies at some vertex 0 of G. Let $i, \tilde{i}$ be the images of 0 in $A, \tilde{A}$ under the maps $\theta, \tilde{\theta}_0$. Now G is freely generated by the subgraph X whose edges are all the elements of the star G_{0*} except e_0 (see Theorem 1, Corollary 2, p.41). Each edge x of X is mapped by θ to an element of the star A_{i*} and this element lifts uniquely to an element $\tilde{x}$ of $\tilde{A}_{\tilde{i}*}$. The graph-map $X \to \tilde{A}$ given by $x \mapsto \tilde{x}$ induces a unique groupoid-map $\tilde{\theta} : G \to \tilde{A}$ which is a lifting of θ and an extension of $\tilde{\theta}_0$. Next, suppose that H is a trivial groupoid. Then $G = N$ is unicursal and the result follows by treating the (simplicial) components of G separately. In the general case, let $T = H \cap N$. Then T is a trivial groupoid with one vertex in each component of N, i.e. a retract of N. We can therefore apply the special case above to obtain a unique groupoid-map $\tilde{\phi} : N \to \tilde{A}$ which lifts the restriction $\phi : N \to A$ of θ and agrees with $\tilde{\theta}_0$ on T. The maps $\tilde{\phi} : N \to \tilde{A}$ and $\tilde{\theta}_0 : H \to \tilde{A}$ now induce a groupoid map $\tilde{\theta} : G = H * N \to \tilde{A}$ with the required properties. ■

COROLLARY. *Let $\alpha : \tilde{A} \to A$ be a covering of groupoids and let $\theta : G \to A$ be a groupoid-map with G connected. Let 0 be a fixed vertex of G and i a fixed vertex of $\tilde{A}$. Then θ lifts to a groupoid-map $\tilde{\theta} : G \to \tilde{A}$ sending 0 to i if and only if $G_{00}\theta \subset \tilde{A}_{ii}\,\alpha$, and in this case $\tilde{\theta}$ is unique.*

Proof. "Only if" is obvious. If $G_{00}\theta \subset \tilde{A}_{ii}\alpha$ then, since α is group-injective (Proposition 31), there is a unique group-map $\tilde{\theta}_0 : G_{00} \to \tilde{A}$ sending 0 to i such that $\tilde{\theta}_0\alpha$ is the restriction of θ to G_{00}. But G_{00} is a full subgroupoid of the connected groupoid G, so $\tilde{\theta}_0$ extends uniquely to a lifting of θ. ■

It is worth noting an alternative proof of Proposition 35 (and hence of Proposition 34 which is a special case). First we may assume that G is connected; H is then also connected since it is a full subgroupoid. Let $\gamma : \tilde{G} \to G$ be the covering of G induced by $\alpha : \tilde{A} \to A$ and $\theta : G \to A$. Since γ is a pull-back of α, there is a unique morphism $\psi : H \to \tilde{G}$ which makes the diagram

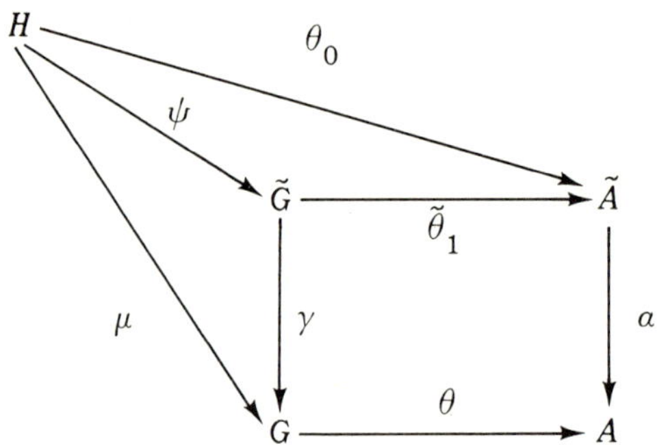

commutative, where $\mu : H \to G$ is the inclusion map. ψ maps H into some component $\tilde{G}_0$ of $\tilde{G}$, and $\psi\gamma = \mu$ is group-surjective. Therefore, since H is not empty, at least one vertex group of $\tilde{G}_0$ maps surjectively to a vertex group of G. By Proposition 31(iv), γ_0 is an isomorphism, and we obtain the required lifting of θ by mapping G first to $\tilde{G}_0$ by γ_0^{-1} and then to $\tilde{A}$ by $\tilde{\theta}_1$ (the pull-back of θ by α). ■

We are now in a position to classify all connected coverings of a given connected groupoid. If H is a subgroup of the groupoid G, say $H \subset G_{ii}$, then a *conjugate* of H is a subgroup of the form $x^{-1}Hx$

where $x \in G_{ij}$ for some j. A *conjugacy class* of subgroups of G is the set of all subgroups conjugate to some given subgroup. For example, the vertex groups of a connected groupoid form a conjugacy class. If $\alpha : \tilde{A} \to A$ is a connected covering of groupoids then the vertex groups of $\tilde{A}$ are mapped injectively into vertex groups of A. Since they are conjugate in $\tilde{A}$ their images are conjugate in A. Further, if the vertex group $\tilde{H}$ of $\tilde{A}$ has image H in A and $x^{-1}Hx$ is any conjugate of H, then x has source in H and lifts uniquely to an edge $\tilde{x}$ of $\tilde{A}$ with source in $\tilde{H}$. Thus $x^{-1}Hx$ is the image of a vertex group $\tilde{x}^{-1}\tilde{H}\tilde{x}$ of $\tilde{A}$, and the image groups form a complete conjugacy class of subgroups of A. We denote this class by $C(\alpha)$, and we write $C(\alpha_1) \leqslant C(\alpha_2)$ if each member of $C(\alpha_1)$ is contained in a member of $C(\alpha_2)$.

PROPOSITION 36. *Let $\alpha_1 : A_1 \to A$ and $\alpha_2 : A_2 \to A$ be connected coverings. Then there is a (covering) morphism $\theta : A_1 \to A_2$ such that $\alpha_1 = \theta\alpha_2$ if and only if $C(\alpha_1) \leqslant C(\alpha_2)$. Hence α_1, α_2 are equivalent coverings if and only if $C(\alpha_1) = C(\alpha_2)$.*

Proof. If $\theta : A_1 \to A_2$ satisfies $\alpha_1 = \theta\alpha_2$ then θ is a covering, by Proposition 33. If H_1 is a vertex group of A_1 then $H_1\theta$ is contained in a vertex group H_2 of A_2 and $H_1\alpha_1 \subset H_2\alpha_2$. Hence $C(\alpha_1) \leqslant C(\alpha_2)$. If θ is an isomorphism then clearly $C(\alpha_1) = C(\alpha_2)$. Conversely, suppose that $C(\alpha_1) \leqslant C(\alpha_2)$. Then any vertex group H_1 of A_1 satisfies $H_1\alpha_1 \subset H_2\alpha_2$ for some vertex group H_2 of A_2 and by Proposition 35, Corollary, the map $\alpha_1 : A_1 \to A$ lifts uniquely to a groupoid-map $\theta : A_1 \to A_2$ sending H_1 into H_2. If $C(\alpha_1) = C(\alpha_2)$ then $H_1\alpha_1 = H_2\alpha_2$ for some H_2 and there are unique morphisms $\theta : A_1 \to A_2$, $\phi : A_2 \to A_1$ such that $\theta\alpha_2 = \alpha_1$, $\phi\alpha_1 = \alpha_2$, θ maps H_1 to H_2 and ϕ maps H_2 to H_1. Clearly θ and ϕ are inverse isomorphisms. ■

THEOREM 7. *Let A be a connected groupoid. Then the mapping $\alpha \mapsto C(\alpha)$ sets up a one-one correspondence between equivalence classes of connected coverings $\alpha : \tilde{A} \to A$ and conjugacy classes of subgroups of A.*

Proof. In view of the last proposition we need only construct, for each conjugacy class C of subgroups of A, a covering α with $C(\alpha) = C$. This basic construction can be described most easily in terms of the corresponding representation of A. Let H be a fixed member of the given conjugacy class C lying at the vertex 0 of A. For any edge a in the star of 0 we define the (right) coset Ha of H to be the set of all edges ha $(h \in H)$. Distinct cosets are disjoint, and $Ha = Hb \Longleftrightarrow ab^{-1} \in H$. If $a \in A_{0i}$ then $Ha \subset A_{0i}$, and we write S_i for the set of distinct cosets contained in A_{0i}. For any $x \in A_{ij}$ and any $Ha \subset A_{0i}$ we have $H(ax) \subset A_{0j}$, so x defines a map $S_i \to S_j$ by $Ha \mapsto Hax$. Clearly this gives us a representation of A on the disjoint sets $S_i (i \in V(A))$, and we denote by $\alpha : \tilde{A} \to A$ the corresponding covering of A. The fibre of α over the vertex i is the set S_i; the fibre over the edge $x \in A_{ij}$ consists of one edge from $Ha \in S_i$ to $Hb \in S_j$ whenever $Hax = Hb$. $\tilde{A}$ is connected since for any two Ha, Hb we have $Ha(a^{-1}b) = Hb$. Now $\tilde{A}$ has one vertex corresponding to the coset H itself. The vertex group at this vertex contains one element lying over each x in A such that $Hx = H$ i.e. $x \in H$. Thus the image of this vertex group in A is precisely H, and it follows that $C(\alpha) = C$. ■

We can also describe this covering groupoid $\tilde{A}$ of A as a concrete groupoid. If $H \in C$ as above and if $Hax = Hb$, then x induces a mapping $Ha \to Hb$ by the rule $ha \mapsto hax$. These *translations* form a groupoid $\mathrm{Tr}(A : H)$ under composition of mappings, the objects of this groupoid being the cosets of H

themselves. If x and y induce the same map $Ha \to Hb$ then $ax = ay$, so $x = y$. It follows that the obvious map from $\tilde{A}$ to $\mathrm{Tr}(A : H)$ is an isomorphism. Or, one can see directly that there is a groupoid-map $\gamma : \mathrm{Tr}(A : H) \to A$ which sends each translation to the unique element of A inducing it. Since each $x \in A_{i*}$ induces a unique translation from Ha to some other coset, for each $Ha \subset A_{*i}$, it is clear that γ is a covering morphism. That γ is connected, with $C(\gamma) = C$, follows as before. ■

If we take C to be the class of trivial subgroups of A then the covering groupoid $\tilde{A}$ has trivial vertex groups, so it is simplicial. It has one vertex for each edge of a star in A. By Proposition 36 this groupoid covers all connected covering groupoids of A and is called the *universal covering groupoid* of A. Example 4 at the beginning of this chapter describes the universal covering groupoid of a group.

It should be noted that the construction of universal covering groupoids is not functorial since if $\theta : A \to B$ is a groupoid-map there will usually be many maps $\tilde{\theta} : \tilde{A} \to \tilde{B}$ between their universal covering groupoids which lie over θ. The induced maps of Proposition 36 are unique only when one specifies the image of one vertex. To rectify this situation one can work in the Category of connected groupoids with base-point (as in [**6**]). This seems a pity since one of the advantages of using groupoids in topology is that they often make base-points unnecessary. However, the only alternative seems to be a more complicated construction on the following lines. The universal covering of A corresponds to the *regular* representation of A by means of the sets $S_i = A_{0i}$ for some fixed vertex 0. It is the choice of 0 that makes the construction not canonical. If instead we take the canonical regular representation of A in which A acts on its costars $S = A_{*i}$ by right multiplication then we obtain a covering groupoid $\hat{A}$ which is no longer connected but has a component isomorphic to the universal cover for each vertex of A. There is now a canonical map $V(A) \to \hat{A}$, and $\hat{A}$ is universal (and functorial) in the Category of coverings with such a lifting of vertices. It is doubtful whether this procedure has any advantage over the use of groupoids with base-point.

Our last theorem on coverings is the crucial result for applications to group theory. It enables us to lift, for example, a free decomposition $A = * A^{\lambda}$ of a groupoid to any covering groupoid. Although the statement looks categorical, its proof is not of the "general nonsense" type. This is one of the few occasions when we need the full force of the solution of the word problem for U_{σ}, and it would be interesting to know if there is any way of by-passing this use of it.

THEOREM 8. *Suppose that* $a : \tilde{A} \to A$ *is a covering map of categories and that*

$$\begin{array}{ccc} \tilde{B} & \xrightarrow{\tilde{\theta}} & \tilde{A} \\ \beta \big\downarrow & & \big\downarrow a \\ B & \xrightarrow{\theta} & A \end{array}$$

is a pull-back square in $\mathcal{C}$. *If* θ *is a universal morphism, then so is* $\tilde{\theta}$. *The same is true for groupoids.*

Proof. We need only treat the case of categories since the terms *covering, universal morphism* and *pull-back* have the same meanings for groupoids as they do when the groupoids are considered as categories.

LEMMA. *Let $\alpha : \tilde{A} \to A$ be a covering of categories, and suppose that $\tilde{a}\alpha = a = a_1 a_2 \dots a_n$ $(n \geqslant 1)$ in A. Then there exist unique elements $\tilde{a}_1, \tilde{a}_2, \dots, \tilde{a}_n$ of $\tilde{A}$ such that $\tilde{a}_r\alpha = a_r$ $(r = 1, 2, \dots, n)$ and $\tilde{a} = \tilde{a}_1\tilde{a}_2 \dots \tilde{a}_n$.*

For $n = 1$ this is trivial. For $n = 2$ we observe that there is a unique $\tilde{a}_1$ covering a_1 whose source is the source of $\tilde{a}$ and a unique $\tilde{a}_2$ covering a_2 whose source is the target of $\tilde{a}_1$. Then $\tilde{a}$ and $\tilde{a}_1\tilde{a}_2$ have the same source and the same image under α, so they are equal. The lemma follows by induction.

To prove the theorem we show that the elements of $\tilde{A}$ are uniquely representable as $\tilde{\sigma}$-reduced words in elements of $\tilde{B}$ *via* the map $\tilde{\theta}$, where $\tilde{\sigma} = V(\tilde{\theta})$. We represent $\tilde{B}$ in standard form as a subcategory of $B \times \tilde{A}$. Its edges are pairs $(b, \tilde{a})$, where $b \in B$, $\tilde{a} \in \tilde{A}$ and $b\theta = \tilde{a}\alpha$. The product $(b_1, \tilde{a}_1)(b_2, \tilde{a}_2)$ is defined in $\tilde{B}$ if and only if *both* products $b_1 b_2$ and $\tilde{a}_1 \tilde{a}_2$ are defined in B, $\tilde{A}$ respectively.

Let $\tilde{a}$ be any non-identity element of $\tilde{A}$ and let $a = \tilde{a}\alpha$. Then a is not an identity of A (Proposition 31 (i)). Now $\theta : B \to A$ is a universal morphism, so by Theorem 4 (p.73), we can write $a = a_1 a_2 \dots a_n$ $(n \geqslant 1)$ where $a_r = b_r\theta$, the b_r are not identities of B, and the products $b_r \cdot b_{r+1}$ are *not* defined in B. By the lemma we can write $\tilde{a} = \tilde{a}_1 \tilde{a}_2 \dots \tilde{a}_n$, where $\tilde{a}_r\alpha = a_r = b_r\theta$, and then the pairs $\tilde{b}_r = (b_r, \tilde{a}_r)$ are in $\tilde{B}$. They are not identities of $\tilde{B}$ (since the b_r are are not identities of B) and the products $\tilde{b}_r \cdot \tilde{b}_{r+1}$ are not defined in $\tilde{B}$ (since $b_r \cdot b_{r+1}$ is not defined in B). Since $\tilde{b}_r\tilde{\theta} = \tilde{a}_r$, $\tilde{a}$ can be written as a $\tilde{\sigma}$-reduced product $\tilde{a} = (\tilde{b}_1\tilde{\theta})(\tilde{b}_2\tilde{\theta})\dots(\tilde{b}_n\tilde{\theta})$. It remains to show that this factorisation is unique and that the identities of $\tilde{A}$ have no such factorisation.

Suppose that we have another such factorisation $\tilde{a} = (\tilde{b}_1'\tilde{\theta})(\tilde{b}_2'\tilde{\theta}) \dots (\tilde{b}_m'\tilde{\theta})$ $(m \geqslant 1)$ where $\tilde{b}_r' = (b_r', \tilde{a}_r')$ is a non-identity of $\tilde{B}$ and $\tilde{b}_r' \cdot \tilde{b}_{r+1}'$ is not defined. Since $\beta : \tilde{B} \to B$ is a covering (Proposition 29) $b_r' = \tilde{b}_r' \beta$ is not an identity of B. Since the products $\tilde{a}_r' \cdot \tilde{a}_{r+1}'$ are defined in A, the products $b_r' \cdot b_{r+1}'$ are *not* defined in B. Now $\tilde{a} = \tilde{a}_1' \tilde{a}_2' \ldots \tilde{a}_m'$, so $a = \tilde{a}\alpha = a_1' a_2' \ldots a_m'$, where $a_r' = \tilde{a}_r' \alpha = b_r' \theta$. By Theorem 4 we must therefore have $m = n$ and $b_r' = b_r (r = 1, 2 \ldots, n)$, since θ is a universal morphism. It follows that $a_r' = a_r$ and therefore, by the lemma, that $\tilde{a}_r' = \tilde{a}_r$. This shows that $\tilde{b}_r' = \tilde{b}_r$. Finally suppose that $(\tilde{b}_1 \tilde{\theta})(\tilde{b}_2 \tilde{\theta}) \ldots (\tilde{b}_n \tilde{\theta})$ is an identity of $\tilde{A}$, where $\tilde{b}_r = (b_r, \tilde{a}_r) \in \tilde{B}$. Then $b_r \theta = \tilde{a}_r \alpha = a_r$, say, and $a_1 a_2 \ldots a_n$ is an identity of A. If the first product is $\tilde{\sigma}$-reduced then, as above, the product $(b_1 \theta)(b_2 \theta) \ldots (b_n \theta)$ is σ-reduced, and this is impossible by Theorem 4 since it represents an identity of A. ■

COROLLARY 1. *If $\alpha : \tilde{A} \to A$ is a covering morphism of categories and $A = * A^\lambda$, where the A^λ are subcategories of A, then $\tilde{A} = * \tilde{A}^\lambda$, where $\tilde{A}^\lambda = A^\lambda \alpha^{-1}$. The same is true for groupoids.*

Proof. Let $B = \coprod A^\lambda$. Then the map $\theta : B \to A$ induced by the inclusion maps $A^\lambda \to A$ is a universal morphism. The induced cover $\tilde{B}$ of B is clearly $\coprod \tilde{A}^\lambda$, where $\tilde{A}^\lambda = A^\lambda \alpha^{-1}$ is the induced cover of A^λ, so by the theorem the map $\coprod \tilde{A}^\lambda \to \tilde{A}$ is universal. Since every vertex of $\tilde{A}$ is in some $\tilde{A}^\lambda$ it follows that $\tilde{A} = * \tilde{A}^\lambda$. ■

COROLLARY 2. *Let $\alpha : \tilde{A} \to A$ be a covering morphism of categories. If A is the free category on a subgraph X, then $\tilde{A}$ is the free category on $\tilde{X} = X\alpha^{-1}$. If A is the free groupoid on X then $\tilde{A}$ is the free groupoid on $\tilde{X}$.*

Proof. Let Y be the absolute free graph on $E(X)$, i.e. Y is the disjoint union of graphs $\bullet \!\longrightarrow\! \bullet$, one for each edge of X. Let B be the free category (groupoid) on Y and let $\theta : B \to A$ be the morphism induced by the canonical graph-map $Y \to X$. Then θ is a universal morphism in both cases. If $\widetilde{Y}$, $\widetilde{B}$ are the induced covers of Y and B, then $\widetilde{Y} \subset \widetilde{B}$ and $\tilde{\theta} : \widetilde{B} \to A$ is a universal morphism. Since $\widetilde{Y}\tilde{\theta} = \widetilde{X}$ and $\widetilde{X}$ contains all vertices of $\widetilde{A}$ it is enough to show that $\widetilde{B}$ is the free category (groupoid) on $\widetilde{Y}$. In other words we need only verify the corollary when X is an absolute free graph, and this reduces immediately to the case when X is the graph $\overset{0}{\bullet} \xrightarrow{\;x\;} \overset{1}{\bullet}$. For groupoids the result is then obvious since A is a simplicial groupoid of type Δ^1, and any covering of Δ^1 consists of disjoint copies of Δ^1 mapped isomorphically to A. For categories there are more possibilities since the covering need not be a co-covering. But it is clear that every edge of $\widetilde{A}$ not an identity must cover x, so $\widetilde{X}$ certainly generates $\widetilde{A}$. Since $\widetilde{X}$ has no paths of length greater than 1, it generates $\widetilde{A}$ freely as a category. ∎

Exercises

1. Prove Corollary 2 above for groupoids by a direct argument using the normal form $x_1^{\epsilon_1} x_2^{\epsilon_2} \ldots x_n^{\epsilon_n}$ for the edges of a free groupoid.

2. (Definition of fibrations). Prove that the following are equivalent for a groupoid-map $\theta : A \to B$:

   (i) θ is star-surjective,

   (ii) θ is a quotient map $A \to A/N$ followed by a covering $A/N \to B$;

   (iii) If

$$\begin{array}{ccc} H \times 0 & \xrightarrow{\quad \alpha \quad} & A \\ {\scriptstyle \mu}\downarrow & & \downarrow{\scriptstyle \theta} \\ H \times \Delta^1 & \xrightarrow[\quad \beta \quad]{} & B \end{array}$$

is a commutative square in $\mathcal{G}$ (with μ the inclusion map) then there is at least one groupoid-map $\gamma : H \times \Delta^1 \to A$ such that $\mu\gamma = \alpha$ and $\gamma\theta = \beta$.

(iv) If

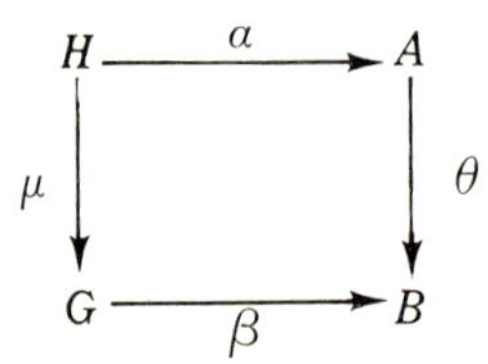

is a commutative square in $\mathcal{G}$, where H is a retract of G and μ the inclusion map, then there is at least one groupoid-map $\gamma : G \to A$ such that $\mu\gamma = \alpha$ and $\gamma\theta = \beta$.

3. Show that any groupoid map ϕ has a factorisation $\phi = \phi_1 \phi_2$, where ϕ_1 is an equivalence of groupoids and ϕ_2 is a fibration (as defined in Exercise 2).

4. Let A be a connected groupoid and construct a graph C as follows. The vertices of C are all subgroups of A. If H, K are subgroups of A, the edges of C from H to K are all triples (H, X, K) where X is a left coset of K which contains a right coset of H. Show that the multiplication $(H, X, K)(K, Y, L) = (H, XY, L)$ makes C a category, and show that this category is equivalent to the Category of connected coverings of A. (XY denotes the set of all products xy which are defined in A, with $x \in X$, $y \in Y$).

CHAPTER 14

Applications to group theory

In this section we use the machinery of groupoids to prove the theorems on groups which are usually known by the names of Nielsen-Schreier, Kurosh and Grushko. For the first two of these we only have to piece together results which we have already proved. We refer the reader to [18], p.28, for the fact that the rank of a free group (i.e. the cardinal of a set of free generators) is an invariant of the group.

THEOREM 9. (Nielsen-Schreier). *Any subgroup H of a free group G is free. If G has finite rank r and H has finite index n in G then H has rank $rn-n+1$.*

Proof. By Theorem 7 (p.110) there is a connected covering $\gamma: \tilde{G} \to G$ and a vertex group H_0 of $\tilde{G}$ such that γ maps H_0 isomorphically to H. For example we may take $\tilde{G} = \mathrm{Tr}(G:H)$. If X is a set of free generators for G, considered as a graph with one vertex, then $\tilde{X} = X\gamma^{-1}$ generates $\tilde{G}$ freely (Theorem 8, Corollary 2). Since $\tilde{G}$ is connected, $\tilde{X}$ is also connected, and it spans $\tilde{G}$. Hence, by Theorem 1, there is a tree $T \subset \tilde{X}$ which spans $\tilde{G}$. Let $Y = \tilde{X} \setminus T$ be the graph obtained from $\tilde{X}$ by deleting the edges of T. Then $\tilde{G} = A * N$, where A and N are the subgroupoids generated by Y and T, respectively, and Y, T generate A, N freely (cf. p.83, Exercise 3). Now N is a simplicial (normal) subgroupoid spanning $\tilde{G}$, so by Theorem 6, we also have a free decomposition $\tilde{G} = H_0 * N$, and there is a unique retraction $\rho : \tilde{G} \to H_0$ with kernel N. Moreover, this retraction induces a universal morphism $A \to H_0$, and since A is free on Y, H_0 is freely generated by the image of Y (Proposition 20). This shows that H is free. (Alternatively, one can use Corollary 2 to Theorem 6). If H has index n in G then $\tilde{G}$ has n vertices and γ is an n-fold covering. Hence if X has r edges, $\tilde{X}$ has rn edges. But T has n-1 edges (Proposition 11, Corollary), so Y has rn-(n-1) edges, and this number is therefore the rank of H. ■

Note. This proof shows that the standard algebraic and topological proofs of the theorem are essentially the same. On the one hand it is just an algebraic model of the proof by covering spaces and fundamental groups. On the other hand, the choice of the tree $T \subseteq \tilde{X}$ is equivalent to the choice of a Schreier transversal of H in G, and one can recover the standard form for the free generators of H by writing down the retraction ρ in terms of the edges of T.

THEOREM 10. (Kurosh). *Let G be a group and suppose that* $G = \underset{\lambda \in \Lambda}{*} G^\lambda$, *where the* G^λ *are subgroups. Then any subgroup H of G has a free decomposition* $H = (\underset{\lambda, \mu}{*} H^{\lambda\mu}) * F$ *with the following properties:*

(i) each $H^{\lambda\mu} (\lambda \in \Lambda, \mu \in M^\lambda)$ *is of the form* $H \cap x_{\lambda\mu} G^\lambda x_{\lambda\mu}^{-1}$, *where, as* μ *varies in* M^λ, $x_{\lambda\mu}$ *runs through a (suitably chosen) set of representatives of the double cosets* $H x G^\lambda$;

(ii) F is a free group; if $l = |\Lambda|$ *is finite, and H has finite index n in G, then F has rank* $ln - m - n + 1$, *where m is the total number of double cosets* $H x G^\lambda$ $(\lambda \in \Lambda)$.

Proof. Let $\gamma : \tilde{G} \to G$ and $H_0 \cong H$ be as in the proof of Theorem 9. Then $\tilde{G} = * \tilde{G}^\lambda$, where $\tilde{G}^\lambda = G^\lambda \gamma^{-1}$. For each λ let $\tilde{G}^{\lambda\mu} (\mu \in M^\lambda)$ be the components of $\tilde{G}^\lambda$. Then $\tilde{G}^\lambda = \ast_\mu \tilde{G}^{\lambda\mu}$ and therefore $\tilde{G} = \ast_{\lambda,\mu} \tilde{G}^{\lambda\mu}$ (Proposition 22). Now $\tilde{G}^{\lambda\mu}$, being connected has a decomposition $\tilde{G}^{\lambda\mu} = K^{\lambda\mu} \ast S^{\lambda\mu}$, where $K^{\lambda\mu}$ is an arbitrary vertex group of $\tilde{G}^{\lambda\mu}$, and $S^{\lambda\mu}$ is a simplicial groupoid spanning $\tilde{G}^{\lambda\mu}$. Hence $\tilde{G} = K \ast S$, where $K = \ast_{\lambda,\mu} K^{\lambda\mu}$ is totally disconnected and $S = \ast_{\lambda,\mu} S^{\lambda\mu}$ is free (since each $S^{\lambda\mu}$ is simplicial, therefore free). Now S spans $\tilde{G}$ because $S^{\lambda\mu}$ spans $\tilde{G}^{\lambda\mu}$ and the $\tilde{G}^{\lambda\mu}$ span $\tilde{G}$. Hence, as in the proof of Theorem 9, we can choose a maximal tree contained in a free generating graph of S and obtain a decomposition $S = A \ast N$ where A is free and N is a simplicial groupoid spanning $\tilde{G}$. Again, there is a unique retraction $\rho : \tilde{G} \to H_0$ with kernel N, and since $\tilde{G} = (K \ast A) \ast N$, ρ induces a universal morphism from $K \ast A = (\ast K^{\lambda\mu}) \ast A$ to H_0. Passing to H by γ we now obtain, by Proposition 23(i), a decomposition $H = (\ast_{\lambda,\mu} H^{\lambda\mu}) \ast F$. Here F is generated by $A\rho\gamma$ and is a free group, by Proposition 20. Since $K^{\lambda\mu}$ is a group its image under $\rho\gamma$ is already a group, so we have $H^{\lambda\mu} = K^{\lambda\mu}\rho\gamma$.

It remains to show that this decomposition of H has the stated properties. The covering $\tilde{G}^\lambda$ of G^λ corresponds to the representation of G^λ (by right multiplication) on the right cosets of H. The components of $\tilde{G}^\lambda$ correspond to the orbits of this representation and are therefore in one-one correspondence with the distinct double cosets $H \mathrm{x} G^\lambda$. For simplicity we take H_0 to be the vertex group of G at the vertex corresponding to the coset H itself. Having chosen a vertex group $K^{\lambda\mu}$ from the component $\tilde{G}^{\lambda\mu}$ of $\tilde{G}^\lambda$, there is a unique edge $\tilde{\mathrm{x}}_{\lambda\mu}$ of N with source in H_0 and target in $K^{\lambda\mu}$, and the image $\mathrm{x}_{\lambda\mu}$ of this edge under γ lies in the corresponding double coset $H \mathrm{x} G^\lambda$. Also, since ρ is the retraction to H_0 with kernel N we have $K^{\lambda\mu}\rho = \tilde{\mathrm{x}}_{\lambda\mu} K^{\lambda\mu} \tilde{\mathrm{x}}_{\lambda\mu}^{-1}$, so $H^{\lambda\mu} = K^{\lambda\mu}\rho\gamma = \mathrm{x}_{\lambda\mu}(K^{\lambda\mu}\gamma)\mathrm{x}_{\lambda\mu}^{-1}$. But $K^{\lambda\mu}\gamma$ is the stabiliser in G^λ of the coset $H\mathrm{x}_{\lambda\mu}$, i.e. $K^{\lambda\mu}\gamma = G^\lambda \cap \mathrm{x}_{\lambda\mu}^{-1} H \mathrm{x}_{\lambda\mu}$, and it follows that $H^{\lambda\mu} = x_{\lambda\mu} G x_{\lambda\mu}^{-1} \cap H$. This proves (i).

In the finite case $\tilde{G}^{\lambda}$ has a finite number m_{λ} of components and n vertices (n the index of H). Each $S^{\lambda\mu}$ is freely generated by a maximal tree in $\tilde{G}^{\lambda\mu}$, so $S^{\lambda} = \underset{\mu}{*} S^{\lambda\mu}$ is freely generated by a maximal circuit-free graph in $\tilde{G}^{\lambda}$. By Proposition 11, such a graph has $n-m_{\lambda}$ edges. Hence $S = \underset{\lambda}{*} S^{\lambda}$ is freely generated by a graph with $ln-m$ edges, where $l = |\Lambda|$, and $m = \sum_{\lambda} m_{\lambda}$ is the total number of components of all the $\tilde{G}^{\lambda}$. It follows that A is freely generated by a graph with $ln-m-(n-1)$ edges (since we have removed a tree with n-1 edges) and F, being of the form $U_{\sigma}(A)$, is a free group with this rank. ■

For the proof of Grushko's theorem we need some auxiliary results on quotients of free products. We say that the normal subgroupoid N of the groupoid A is *adapted to the free decomposition* $A = *A^{\lambda}$ (where $A^{\lambda} \subset A$) if the quotient groupoid $B = A/N$ is the free product of the subgroupoids B^{λ} generated by the images of the A^{λ}. For example, this is the case, by Proposition 27, Corollary, if N is generated as normal subgroupoid by its intersections with the A^{λ}. For groups the converse of this statement is also true; for groupoids, unfortunately, it is not true, and we must examine the situation more closely.

All free products in the rest of this section are free products of subgroupoids.

PROPOSITION 37. *Let $A = *A^{\lambda}$ and let M, N be normal subgroupoids of A with $M \subset N$. Assume that M is adapted to the decomposition $A = *A^{\lambda}$, and let $B = *B^{\lambda}$ be the corresponding decomposition of $B = A/M$. Then N is adapted to $*A^{\lambda}$ if and only if N/M is adapted to $*B^{\lambda}$.*

Proof. Let $\mu : A \to A/M$ and $\nu : A \to A/N$ be the quotient maps. The induced map $\pi : A/M \to A/N$ is a quotient map with kernel M/N. The subgroupoids of A/N generated by the $B^\lambda \pi$ are the same as the subgroupoids generated by the $A^\lambda \nu$, and the result follows immediately. ■

PROPOSITION 38. *Let $\{N_i\}_{i \in I}$ be a family of normal subgroupoids of A, and let N be the normal subgroupoid generated by their union. If each N_i is adapted to the decomposition $A = * A^\lambda$, then so is N.*

Proof. This is another example of Proposition 18 (right limits commute with right limits). Write $B_i = A/N_i$, $B = A/N$, with quotient maps $\pi_i : A \to B_i$, $\pi : A \to B$. Then B is the right limit of the diagram with objects A, B_i and morphisms π_i $(i \in I)$. Each of A, B_i is a free product indexed by $\lambda \in \Lambda$, so can be expressed as the right limit of an appropriate diagram (over the same graph D in each case). The maps π_i respect the free decompositions, so give rise to maps of D-diagrams, and the rest is tedious checking.

More directly, writing B_i^λ and B^λ for the groupoids generated by the images of A^λ in B_i and B, we have $B_i = * B_i^\lambda$ for all i and we wish to prove $B = * B^\lambda$. Since $N_i \subset N$, there is a unique morphism $\phi_i : B_i \to B$ such that $\pi_i \phi_i = \pi$; and π_i, ϕ_i, π induce morphisms

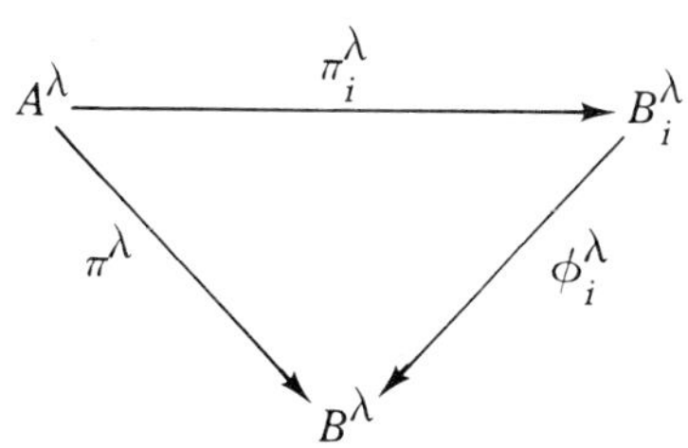

such that $\pi_i^\lambda \phi_i^\lambda = \pi^\lambda$ for all i, λ. Suppose that $\beta^\lambda : B^\lambda \to C$ are morphisms which agree on identities. Then the morphisms

$\alpha^\lambda = \pi^\lambda\beta^\lambda : A^\lambda \to C$ agree on identities and therefore induce a morphisms $\alpha : A \to C$. Also, for each $i \in I$, the morphisms $\beta_i^\lambda = \phi_i^\lambda \beta^\lambda : B_i^\lambda \to C$ agree on identities and induce a morphism $\beta_i : B_i \to C$. Now, for each λ, the restriction of $\pi_i \beta_i$ to A^λ is $\pi_i^\lambda \beta_i^\lambda = \pi^\lambda \beta^\lambda = \alpha^\lambda$, so $\pi_i \beta_i = \alpha$. In particular Ker $\alpha \supset$ Ker $\pi_i = N_i$, and this for all $i \in I$. Hence Ker $\alpha \supset N$ and there is a unique morphism $\beta : B \to C$ such that $\pi\beta = \alpha$. This β agrees with β^λ on $A^\lambda \pi^\lambda$, and $A^\lambda \pi^\lambda$ generates B^λ, so the restriction of β to B^λ is β^λ. The uniqueness of β subject to this last condition is obvious since any such β must satisfy $\pi\beta = \alpha$. This proves that $B = *B^\lambda$. ■

PROPOSITION 39. *Let $A = * A^\lambda$ and let N be a normal subgroupoid of A adapted to this decomposition. If each $N \cap A^\lambda$ is totally disconnected then so is N.*

Proof. Every non-identity a of A is uniquely expressible as a reduced word $a = a_1 a_2 \dots a_n$ ($n \geqslant 1$, $a_i \in A^{\lambda_i}$, $\lambda_i \neq \lambda_{i+1}$, a_i not an identity). We call n the length of a; identities have length zero. Suppose that N is not totally disconnected, and let a be an element of N joining distinct vertices and of minimal length subject to these conditions. If $a = a_1 a_2 \dots a_n$ is its standard form and if $b_i = a_i\pi$, where $\pi : A \to B = A/N$ is the quotient map, then $b_1 b_2 \dots b_n$ is an identity of B. But $B = * B^\lambda$, where B^λ is generated by $A^\lambda\pi$, and $b_i \in B^{\lambda_i}$ ($\lambda_i \neq \lambda_{i+1}$). By Theorem 5(p.81) it follows that some b_i is an identity, i.e. $a_i \in N \cap A^{\lambda_i}$ for some i. But $N \cap A^{\lambda_i}$ is totally disconnected, so this a_i lies in a vertex group of N. Hence the product $a' = a_1 a_2 \dots a_{i-1} a_{i+1} \dots a_n$ is defined in A, and the element a' has length at most n-1. However, a' has the same source and target as a, and $a' \in N$ since $a'\pi = b_1 b_2 \dots b_{i-1} b_{i+1} \dots b_n$ is still an identity of B. This contradiction proves the proposition. ■

THEOREM 11. *Let* $A = * A^{\lambda}$ *and let* N *be a normal subgroupoid adapted to this decomposition. Then there is a unicursal normal subgroupoid* $N_0 \subset N$ *which spans* N *and is also adapted to the decomposition.*

Proof. Let $\mathfrak{N}$ be the set of all unicursal normal subgroupoids contained in N which are adapted to the decomposition. $\mathfrak{N}$ is not empty since it contains the trivial normal subgroupoid T of A. Note that a unicursal subgroupoid is normal if and only if it contains T. If $\{N_i\}_{i \in I}$ is a chain in $\mathfrak{N}$ (ordered by inclusion) then $\bigcup_{i \in I} N_i$ is clearly unicursal, normal and contained in N. Since it is generated by the N_i, which are all adapted to the decomposition, Proposition 38 tells us that $\bigcup N_i$ is also adapted. Thus $\mathfrak{N}$ is inductively ordered by inclusion and contains a maximal member N_0, by Zorn's lemma. It remains to show that N_0 spans N, i.e. that $\overline{N} = N/N_0$ is totally disconnected. Let $\overline{A} = A/N_0$, with quotient map $\pi : A \to \overline{A}$. Then $\overline{A} = * \overline{A}^{\lambda}$, where $\overline{A}^{\lambda}$ is generated by $A^{\lambda}\pi$. By Proposition 37, $\overline{N}$ is adapted to this decomposition. Suppose that $\overline{N}$ is *not* totally disconnected. Then, by Proposition 39, some $\overline{A}^{\lambda} \cap \overline{N}$ is not totally disconnected, so there is an $\overline{x} \in \overline{A}^{\lambda} \cap \overline{N}$ joining distinct vertices. The normal subgroupoid $\overline{X}$ generated by $\overline{x}$ consists of $\overline{x}, \overline{x}^{-1}$ and all the identities of $\overline{A}$. It is clearly adapted to the decomposition $\overline{A} = * \overline{A}^{\lambda}$ (see Proposition 27, Corollary), and is contained in $\overline{N}$. Now $\overline{X} = X/N_0$, where $X = \overline{X}\pi^{-1}$ is a normal subgroupoid of A contained in N. By Proposition 37, X is adapted to the decomposition $A = * A^{\lambda}$. Also X is unicursal since X/N_0 and N_0 are both unicursal. Thus $X \in \mathfrak{N}$. But X/N_0 is not trivial, so X contains N_0 properly, and this contradiction proves the theorem. ∎

THEOREM 12. *Let* G, B *be groups with free decompositions* $G = * G^{\lambda}$, $B = * B^{\lambda}$ $(\lambda \in \Lambda)$, *and let* $\psi : G \to B$ *be a group* *homomorphism such that* $G^\lambda\psi = B^\lambda$ *for all* $\lambda \in \Lambda$. *If* H *is any subgroup of* G *such that* $H\psi = B$, *then* H *has a decomposition* $H = * H^\lambda$ *such that* $H^\lambda\psi = B^\lambda$ *for all* λ.

Proof. Let $A = \mathrm{Tr}(G:H)$ and let $\alpha : A \to G$ be the standard covering morphism. Then A is connected and has a vertex group $H_0 = A_{00}$ which maps isomorphically to H under α. Consider the groupoid-map $\theta = \alpha\psi : A \to B$. If i, j are vertices of A then there exist edges $x \in A_{0i}, y \in A_{0j}$. Since $H_0\theta = H\psi = B$ (a group) we can find, for each $b \in B$ an element $h \in H_0$ such that $h\theta = (x\theta)\, b(y\theta)^{-1}$. Then $a = x^{-1}hy \in A_{ij}$ and $a\theta = b$, so θ is piece-wise surjective. Since θ is obviously vertex-surjective it follows that it is a quotient map, by Proposition 25. Thus $B \cong A/N$, where $N = \mathrm{Ker}\ \theta$, and since B is a group, N spans A. Also, by Theorem 8, Corollary 1, $A = * A^\lambda$, where $A^\lambda = G^\lambda\alpha$. Since $A^\lambda\theta = G^\lambda\psi = B^\lambda$, N is adapted to this decomposition of A. Hence, by Theorem 11, there is a unicursal normal subgroupoid N_0 spanning N which is also adapted to the decomposition. N_0 spans A (since N does), so by Theorem 6 (p.92), $A = H_0 * N_0$, and there is a unique retraction $\rho : A \to H_0$ with kernel N_0. Since ρ is a quotient map (Proposition 28) it follows that $H_0 = * H_0^\lambda$, where H_0^λ is generated by $A^\lambda\rho$. Hence $H = * H^\lambda$, where $H^\lambda = H_0^\lambda\alpha$. Finally, since $\mathrm{Ker}\ \rho = N_0 \subset N$ there is a unique morphism $\theta^* : H_0 \to B$ such that $\rho\theta^* = \theta$. Hence $H^\lambda\psi = H_0^\lambda\theta = H_0^\lambda\rho\theta^* = H_0^\lambda\theta^*$ (since ρ restricted to H_0 is the identity), and $H^\lambda\psi$ is therefore generated by $(A^\lambda\rho)\,\theta^* = A^\lambda\theta = B^\lambda$. But B^λ is a group, so $H^\lambda\psi = B^\lambda$ as required. ■

COROLLARY. (Grushko's Theorem). *Let* B *be a group with a free decomposition* $B = * B^\lambda$, *and let* F *be a free group. If* $\phi : F \to B$ *is a surjective group homomorphism, then* F *has a decomposition* $F = * F^\lambda$ *such that* $F^\lambda\phi = B^\lambda$.

Proof. Let X be a set of free generators for F, and let the image $x\phi$ of each $x \in X$ be written in some fashion as $x\phi = b_{x,1} b_{x,1} \cdots b_{x,r}$, where $r = r(x) \geqslant 1$ and each $b_{x,i}$ lies in some B^λ. This is certainly possible, and we take a fixed representation of this form for each x. We also fix, for each pair x, i a $\lambda = \lambda(x,i)$ such that $b_{x,i} \in B^\lambda$. (This is necessary since some of the $b_{x,i}$ may be identities and lie in several B^λ). Now let Y be a set whose members are distinct symbols $y_{x,i}$ for $x \in X$ and $i = 1,2,\ldots,r(x)$. Then $Y = \coprod Y^\lambda$, where $Y^\lambda = \{y_{x,i} \mid \lambda(x,i) = \lambda\}$. The free group G on the set Y has a free decomposition $G = * G^\lambda$, where G^λ is free on Y^λ. The unique homomorphism $\psi : G \to B$ defined by $y_{x,i} \mapsto b_{x,i}$ maps G^λ into B^λ. Also, the unique homomorphism $\sigma : F \to G$ defined by $x \mapsto y_{x,1} y_{x,2} \cdots y_{x,r}$ is an injection, since the $y_{x,i}$ are all distinct free generators of G. For each $x \in X$ we have $x\sigma\psi = b_{x,1} b_{x,2} \cdots b_{x,r} = x\phi$, so $\sigma\psi = \phi$. Hence the subgroup $H = F\sigma$ of G satisfies $H\psi = F\phi = B$. By the theorem $H = * H^\lambda$, where $H^\lambda\psi = B^\lambda$, and it follows that $F = * F^\lambda$, $F^\lambda\phi = B^\lambda$, where $F^\lambda = H^\lambda\sigma^{-1}$. ■

Problem. Can one strengthen the conclusion of Theorem 12 so that in the special case when B is trivial it reduces to the Kurosh subgroup theorem? A suitable conjecture would be that under the hypotheses of Theorem 12, H has a free decomposition $H = * H^\lambda$ such that (i) $H^\lambda\phi = B^\lambda$ and (ii) each H^λ has a decomposition $H^\lambda = (\underset{\mu}{*} H^{\lambda\mu}) * F^\lambda$, where F^λ is free, $H^{\lambda\mu} = H \cap x_{\lambda\mu} G^\lambda x_{\lambda\mu}{}^{-1}$, and for fixed λ the $x_{\lambda\mu}$ are a set of representatives of the double cosets $H x G^\lambda$. This conjecture is true when H has finite index in G, as one can show by the following line of argument. The covering A of G in the proof of Theorem 12 is mapped to H_0 by a retraction ρ whose kernel N_0 is adapted to the decomposition $A = * A^\lambda$. Suppose that A has a finite number $n \geqslant 2$ of vertices (n is the index of H in G). By

Proposition 39, N_0 contains an edge x of some A^λ joining distinct vertices. We may retract A to a full subgroupoid A_1 with n-1 vertices using as kernel the normal subgroupoid X generated by x. By Proposition 27, Corollary, $A_1 = * A_1^\lambda$, where A_1^λ is generated by the image of A^λ. Also, if $x \in A^\lambda$, then the induced map $A^\mu \to A_1^\mu$ is *universal* if $\mu \neq \lambda$ and a *retraction* if $\mu = \lambda$. Now N_0/X is adapted to this decomposition, so we can do the same trick again. Thus we obtain successive retractions $A = A_0 \to A_1 \to \ldots \to A_{n-1} = H_0$ with $A_i = \underset{\lambda}{*} A_i^\lambda$ and induced maps $A_i^\lambda \to A_{i+1}^\lambda$ either universal or retractions. The composite of these maps is ρ, so A_{n-1}^λ is the H_0^λ of the theorem. Now let $K^{\lambda\mu}$ be chosen as in the Kurosh theorem, i.e. one vertex group from each component of each A^λ. Then by induction one can prove that any subgroupoid D_i^λ of A_i^λ consisting of one vertex group from each component has the form $(\underset{\mu}{*} K_i^{\lambda\mu}) * F_i^\lambda$ where F_i^λ is free,and $K^{\lambda\mu}$ is a suitable conjugate in A of $K^{\lambda\mu}$. This is true when $i = 0$ since all vertex groups of a component of A^λ are conjugate in A^λ. The hypothesis clearly carries over from A_i^λ to A_{i+1}^λ if the induced map $A_i^\lambda \to A_{i+1}^\lambda$ is a retraction since A_{i+1}^λ is then a full subgroupoid of A_i^λ meeting each component. Suppose on the other hand that the map $A_i^\lambda \to A_{i+1}^\lambda$ is universal and assume that D_i^λ (a totally disconnected retract of A_i^λ) has the stated form. Then $A_i^\lambda = D_i^\lambda * P$ where P is free. The map $A_i^\lambda \to A_{i+1}^\lambda$, being induced by a retraction $A_i \to A_{i+1}$, sends any subgroup of A_i^λ to a conjugate subgroup. Since the map is universal it preserves all free decompositions; hence $A_{i+1}^\lambda = \overline{D}_i^\lambda * Q$, where $\overline{D}_i^\lambda$ is a free product of a set of conjugates of the $K^{\lambda\mu}$ and a free group, while Q is a free groupoid (of the form $U_\sigma(P)$). It follows that Q spans A_{i+1}^λ and is of the form $Q = R * S$ where R is a totally disconnected free groupoid and S is unicursal spanning A_{i+1}^λ. Now take any subgroupoid D_{i+1}^λ

of A_{i+1}^{λ} consisting of one vertex group from each component. Then there is a unique retraction $A_{i+1}^{\lambda} \to D_{i+1}^{\lambda}$ with kernel S. Since $A_{i+1}^{\lambda} = \bar{D}_i^{\lambda} * Q = (\bar{D}_i^{\lambda} * R) * S$, the induced map $(\bar{D}_i^{\lambda} * R) \to D_{i+1}^{\lambda}$ is universal (Theorem 6). Again it sends subgroups to conjugates of themselves, and we deduce that D_{i+1}^{λ} has the required form. This completes the induction, and it follows that $H_0 = * H_0^{\lambda}$, where $H_0^{\lambda} = A_{n-1}^{\lambda} = (\underset{\mu}{*} H_0^{\lambda\mu}) * F_0^{\lambda}$, where F_0^{λ} is free and $H_0^{\lambda\mu}$ is some conjugate in A of $K^{\lambda\mu}$. The corresponding decomposition of H then has all the stated properties.

A similar inductive proof for the general case would require a limit argument, and this seems rather elusive.

Note added in proof. Some results related to this problem have been proved recently by E. T. Ordman [**26**], [**27**].

Exercises

1. Prove that any subgroupoid of a free groupoid is free.

2. Let $A = * A^{\lambda}$ and $B = * B^{\lambda}$ be groupoids and let $\theta : A \to B$ be a morphism such that $A^{\lambda}\theta \subset B^{\lambda}$. Prove that Ker θ is adapted to the decomposition $A = * A^{\lambda}$.

3. Prove that if a totally disconnected normal subgroupoid N of A is adapted to a decomposition $A = * A^{\lambda}$, then N is generated as normal subgroupoid by all the $N \cap A^{\lambda}$.

4. Let $A = * A^{\lambda}$, let $\theta : A \to B$ be a universal morphism of groupoids, and let $B = * B^{\lambda}$ be the corresponding decomposition of B. Prove that if M is a normal subgroupoid of A, and N is the normal subgroupoid of B generated by $M\theta$, then M is adapted to $*A^{\lambda}$ if and only if N is adapted to $*B^{\lambda}$.

CHAPTER 15

Coverings of right limits

Let $\mathbf{A}$ be a diagram in $\mathcal{C}$ consisting of categories $A^{\lambda}(\lambda \in \Lambda)$ and morphisms $\phi^{\sigma}(\sigma \in \Sigma)$. Let $A = \varinjlim \mathbf{A}$ with canonical morphisms $\alpha^{\lambda} : A^{\lambda} \to A$. If $\gamma : \tilde{A} \to A$ is a covering morphism then we obtain, *via* the maps α^{λ}, induced coverings $\gamma^{\lambda} : \tilde{A}^{\lambda} \to A^{\lambda}$ and maps $\tilde{\alpha}^{\lambda} : \tilde{A}^{\lambda} \to \tilde{A}$. For any $\phi^{\sigma} : A^{\lambda} \to A^{\mu}$ such that $\phi^{\sigma}\alpha^{\mu} = \alpha^{\lambda}$, there is a unique $\tilde{\phi}^{\sigma} : \tilde{A}^{\lambda} \to \tilde{A}^{\mu}$ such that $\tilde{\phi}^{\sigma}\tilde{\alpha}^{\mu} = \tilde{\alpha}^{\lambda}$ and $\tilde{\phi}^{\sigma}\gamma^{\mu} = \gamma^{\lambda}\phi^{\sigma}$. We therefore obtain a diagram $\tilde{\mathbf{A}}$ and a diagram-map from $\tilde{\mathbf{A}}$ to the trivial diagram $\Gamma(\tilde{A})$. We aim to prove

THEOREM 13. *If $\gamma : \tilde{A} \to A$ is a covering of categories and $A = \varinjlim \mathbf{A}$ in $\mathcal{C}$, then $\tilde{A} = \varinjlim \tilde{\mathbf{A}}$, where $\tilde{\mathbf{A}}$ is the diagram of induced covers. The same is true in $\mathcal{G}$.*

Note. The forgetful functor $F : \mathcal{G} \to \mathcal{C}$ has a right adjoint, namely, the functor which assigns to each category its groupoid of invertible elements. Hence F preserves right limits, by Proposition 15. We already know that F preserves left limits, in particular pull-backs, and F also sends coverings to coverings. Hence we need only prove Theorem 13 for categories.

There is another convenient formulation of this result. If Z is any object of a Category $\mathcal{K}$, we can form the Category $\mathcal{K}_Z$ of $\mathcal{K}$-objects and morphisms over Z as follows. The objects of $\mathcal{K}_Z$ are $\mathcal{K}$-morphisms $\beta : B \to Z$, for arbitrary B in $\mathcal{K}$, and if $\beta' : B' \to Z$ is another such then the $\mathcal{K}_Z$-morphisms from β to β' are commutative triangles

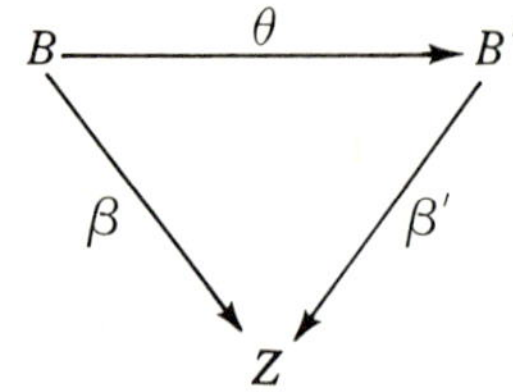

in $\mathcal{K}$. A D-diagram in $\mathcal{K}_Z$ is essentially the same thing as a D-diagram $\mathbf{A}$ in $\mathcal{K}$ together with a diagram-map from $\mathbf{A}$ to the trivial diagram $\Gamma(Z)$. If $\mathcal{K}$ is right complete then $\varinjlim \mathbf{A}$ exists in $\mathcal{K}$, and the induced map $\varinjlim \mathbf{A} \to Z$ is clearly the right limit of the corresponding diagram in $\mathcal{K}_Z$. Thus $\mathcal{K}_Z$ is also right complete.

Now let $\zeta : Y \to Z$ be a fixed $\mathcal{K}$-morphism and suppose that $\mathcal{K}$ admits pull-backs. Then for each object $\beta : B \to Z$ of $\mathcal{K}_Z$ we have a pull-back square

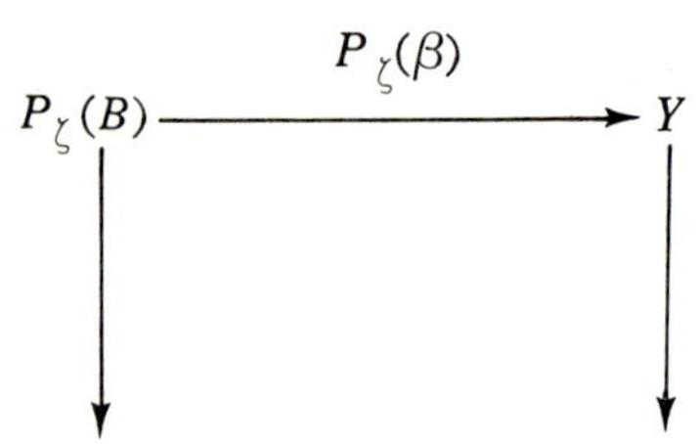

which gives us an object $P_\zeta(\beta)$ of $\mathcal{K}_Y$. Any $\mathcal{K}_Z$-morphism lifts uniquely to a $\mathcal{K}_Y$-morphism, and we obtain a *pull-back functor* $P_\zeta : \mathcal{K}_Z \to \mathcal{K}_Y$. It is an easy exercise to show that Theorem 13 is equivalent to the following:

THEOREM 13′. *If $\zeta : \tilde{Z} \to Z$ is a covering morphism in $\mathcal{C}$ then the pull-back functor $P_\zeta : \mathcal{C}_Z \to \mathcal{C}_{\tilde{Z}}$ preserves right limits.*

Now a functor $F : \mathcal{K} \to \mathcal{L}$, where $\mathcal{K}$ is right complete, preserves right limits if and only if it preserves coproducts and difference cokernels. This is clear from the proof of Proposition 16 (or rather the dual of this proof) since all the constructions involved are preserved by such a functor. Hence we need only prove that P_ζ in Theorem 13′ preserves coproducts and difference cokernels, and this is again equivalent to proving Theorem 13 in these special cases. For coproducts the result is trivial since any cover of a disjoint union of categories A^λ is the disjoint union of the induced covers of the A^λ. Theorem 13 is therefore reduced to the following:

PROPOSITION 40. *In the diagram of categories*

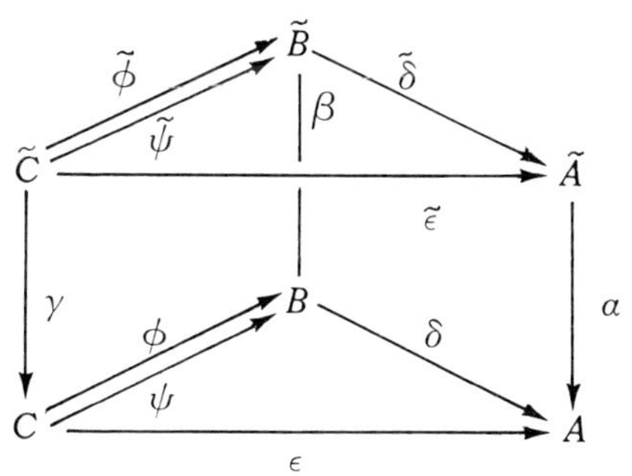

let δ be the difference cokernel of ϕ and ψ, with $\phi\delta = \psi\delta = \epsilon$, and let a be a covering morphism. Let $(\tilde{\delta}, \beta)$ be the pull-back of (a, δ) and let $(\tilde{\epsilon}, \gamma)$ be the pull-back of (a, ϵ). Finally, let $\tilde{\phi}, \tilde{\psi}$ be the unique morphisms satisfying $\tilde{\phi}\tilde{\delta} = \tilde{\psi}\tilde{\delta} = \tilde{\epsilon}$, $\tilde{\phi}\beta = \gamma\phi$ and $\tilde{\psi}\beta = \gamma\psi$. Then $\tilde{\delta}$ is the difference cokernel of $\tilde{\phi}$ and $\tilde{\psi}$.

Proof. First suppose that $V(\phi) = V(\psi) : K \to I$. We recall from Ch. 9 the following facts which characterise δ in this case: (i) $V(A) = I$ and δ is an I-morphism (that is $V(\delta) = 1_I$); (ii) δ is surjective; (iii) for $x, y \in B$, $x\delta = y\delta$ if and only if $x \equiv y$, where $\equiv$ is the equivalence relation on the edges of B generated by all pairs of the form $(b_1(c\phi)b_2, b_1(c\psi)b_2)$, where $c \in C$ and $b_1, b_2 \in B$. If $V(\tilde{A}) = \tilde{I}$, then $V(\tilde{B})$ can be identified with $\tilde{I}$, and with this identification $\tilde{\delta}$ is an $\tilde{I}$-morphism. Clearly $\tilde{\delta}$ is surjective, so it remains to prove the analogue of (iii) for $\tilde{\delta}$. We write $x \sim y$ if x and y are of the form $b_1(c\phi)b_2$ and $b_1(c\psi)b_2$ respectively, and we also use the symbols $\sim$ and $\equiv$ for the corresponding relations in $\tilde{B}$ defined in terms of $\tilde{\phi}$ and $\tilde{\psi}$. Since $\tilde{\phi}\tilde{\delta} = \tilde{\psi}\tilde{\delta}$ by construction, $\tilde{x} \sim \tilde{y}$ implies $\tilde{x}\tilde{\delta} = \tilde{y}\tilde{\delta}$. Hence $\tilde{x} \equiv \tilde{y}$ implies $\tilde{x}\tilde{\delta} = \tilde{y}\tilde{\delta}$. Conversely, suppose that $\tilde{x}, \tilde{y} \in \tilde{B}$ and $\tilde{x}\tilde{\delta} = \tilde{y}\tilde{\delta} = \tilde{a}$, say. Then, taking $\tilde{B} \subset B \times \tilde{A}$ and $\tilde{C} \subset C \times \tilde{A}$ as usual, we have $\tilde{x} = (x, \tilde{a})$, $\tilde{y} = (y, \tilde{a})$, where $x\delta = y\delta = \tilde{a}\alpha = a$, say. By (iii) above, there exist edges $x = x_0, x_1, \ldots, x_n = y$ in B such that for $i = 1, 2, \ldots, n$, either $x_i \sim x_{i-1}$ or $x_{i-1} \sim x_i$. Since $x_i\delta = a = \tilde{a}\alpha$ for all i, there are corresponding edges $\tilde{x}_i = (x_i, \tilde{a})$ in $\tilde{B}$, and we need only show that $\tilde{x}_i \sim \tilde{x}_{i-1}$ or $\tilde{x}_{i-1} \sim \tilde{x}_i$. Thus it suffices to assume that $x \sim y$ and deduce that $\tilde{x} \sim \tilde{y}$. Suppose that $x = b_1(c\phi)b_2$ and $y = b_1(c\psi)b_2$, and let $b_1\delta = a_1$, $b_2\delta = a_2$, $c\phi\delta = c\psi\delta = a_0$. Then $a_1a_0a_2 = x\delta = y\delta = a$. Since α is a covering, this factorisation of a lifts (uniquely) to a factorisation $\tilde{a} = \tilde{a}_1\,\tilde{a}_0\,\tilde{a}_2$, with $\tilde{a}_i\,\alpha = a_i$ for $i = 0,1,2$ (see the lemma on p.113). This gives us elements $\tilde{b}_1 = (b_1, \tilde{a}_1)$, $b_2 = (b_2, \tilde{a}_2)$ in $\tilde{B}$ and $\tilde{c} = (c, \tilde{a}_0)$ in $\tilde{C}$ such that $\tilde{b}_1(\tilde{c}\tilde{\phi})\tilde{b}_2 = \tilde{x}$ and $\tilde{b}_1(\tilde{c}\tilde{\psi})\tilde{b}_2 = \tilde{y}$. Hence $\tilde{x} \sim \tilde{y}$ as required.

In the general case, let $\sigma = V(\delta)$ and write $B_0 = U_\sigma(B)$, with canonical map $\xi : B \to B_0$. Then we have a diagram

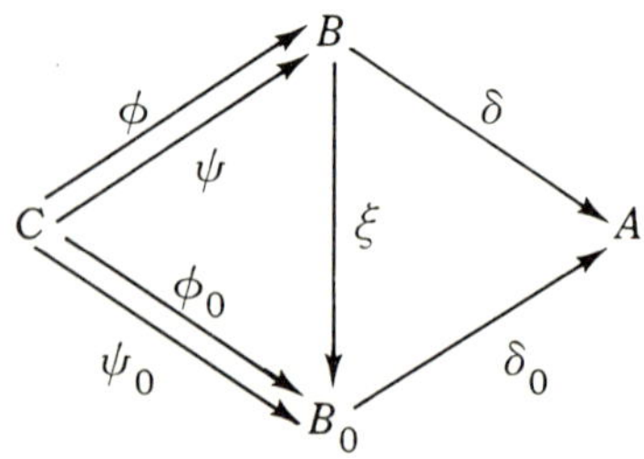

with $\phi_0 = \phi\xi$, $\psi_0 = \psi\xi$ and $\xi\delta_0 = \delta$. The pull-back functor P_α gives a corresponding diagram

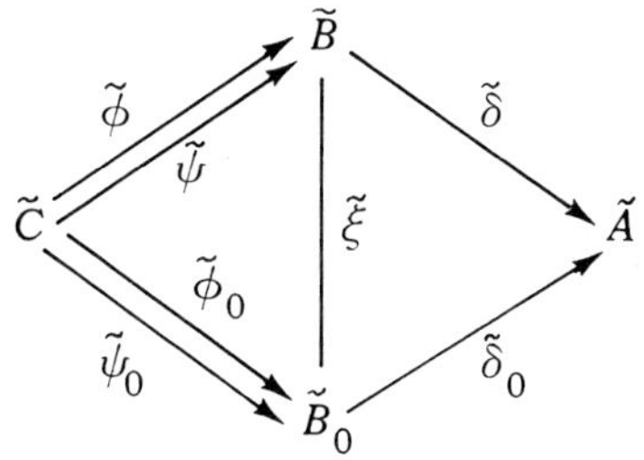

which covers the first diagram in the sense that the coverings α, β, γ and the induced covering $\beta_0 : \tilde{B}_0 \to B_0$ form a morphism of diagrams. Now δ_0 is vertex-bijective and is the difference cokernel of ϕ_0 and ψ_0. By the special case already dealt with, it follows that $\tilde{\delta}_0$ is the difference cokernel of $\tilde{\phi}_0$ and $\tilde{\psi}_0$. It is easy to check also that

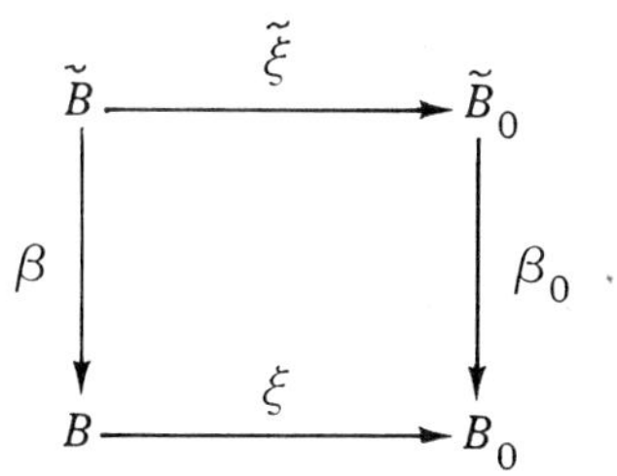

is a pull-back square. Since ξ is a universal morphism, it follows from Theorem 8(p.112) that $\tilde{\xi}$ is a universal morphism. Finally, since $V(\xi) = V(\delta)$ is the difference cokernel (in $\mathcal{S}$) of $V(\phi)$ and $V(\psi)$, an easy argument shows that $V(\tilde{\xi}) = V(\tilde{\delta})$ is the difference cokernel of $V(\tilde{\phi})$ and $V(\tilde{\psi})$. These facts imply that $\tilde{\delta}$ is the difference cokernel of $\tilde{\phi}$ and $\tilde{\psi}$. For if $\tilde{\phi}\theta = \tilde{\psi}\theta$ then $V(\theta)$ factorises through $V(\tilde{\xi})$, so $\theta = \tilde{\xi}\theta_1$ uniquely. θ_1 satisfies $\tilde{\phi}_0\theta_1 = \tilde{\psi}_0\theta_1$, so $\theta_1 = \tilde{\delta}_0\theta_2$ uniquely. Hence $\theta = \tilde{\delta}\theta_2$ and θ_2 is clearly unique with respect to this property. This proves the proposition and hence Theorems 13 and 13′. ■

Example 1. Free products can be expressed as right limits, so Theorem 13 includes Corollary 1 of Theorem 8. More generally, call the category (or groupoid) A the *generalised free product* of subcategories (subgroupoids) A^λ if, for arbitrary morphisms $\theta^\lambda : A^\lambda \to B$ such that θ^λ and θ^μ agree on $A^{\lambda\mu} = A^\lambda \cap A^\mu$ for all λ, μ, there is a unique morphism $\theta : A \to B$ whose restriction to A^λ is θ^λ for all λ. This is equivalent to saying that A is the right limit of the diagram with objects all A^λ, $A^{\lambda\mu}$ and morphisms all the inclusions $A^{\lambda\mu} \to A$, $A^{\lambda\mu} \to A^\mu$. We write $A = \overline{*} A^\lambda$ in this situation. If $a : \tilde{A} \to A$ is a covering then the induced covers $\tilde{A}^\lambda$ of A^λ and $\tilde{A}^{\lambda\mu}$ of $A^{\lambda\mu}$ are all subcategories (subgroupoids) of $\tilde{A}$ with $\tilde{A}^{\lambda\mu} = \tilde{A}^\lambda \cap \tilde{A}^\mu$, and Theorem 13 implies that $\tilde{A} = \overline{*} \tilde{A}^\lambda$. (Generalised free products are also called free products with amalgamations, but there is some ambiguity since the latter term is also used to describe arbitrary right limits A of diagrams consisting of categories (or groupoids) A^λ, $A^{\lambda\mu}$ and injective morphisms $A^{\lambda\mu} \to A^\lambda$, $A^{\lambda\mu} \to A^\mu$. In this situation the A^λ are not necessarily embedded in A.

Example 2. Presentations of categories and groupoids are examples of right limits, so presentations can be lifted to coverings. A little care is needed in formulating the result correctly since the construction is different in the two cases. First suppose that $A = \text{cat}(X; R)$. Then we have a diagram $F\mathcal{C}(R) \underset{\psi}{\overset{\phi}{\rightrightarrows}} \vec{P}(X) \xrightarrow{\delta} A$, where δ is the difference cokernel of ϕ and ψ. If $a : \tilde{A} \to A$ is both a covering and a co-covering then the induced cover of $F\mathcal{C}(R)$ is again an absolute free category, while the induced cover of $\vec{P}(X)$ is $\vec{P}(\tilde{X})$ ($\tilde{X}$ the induced cover of X). Thus we obtain a presentation $\tilde{A} = \text{cat}(\tilde{X}; \tilde{R})$, where R consists of pairs $(\tilde{r}_1, \tilde{r}_2)$ of edges of $\vec{P}(\tilde{X})$ which, in the standard model for induced covers, are of the form $\tilde{r}_1 = (r_1, \tilde{a}), \tilde{r}_2 = (r_2, \tilde{a})$ with $(r_1, r_2) \in R$ and $\tilde{a}a = r_1\delta = r_2\delta$. If a is a covering but not a co-covering the induced cover of $F_{\mathcal{C}}(R)$ may contain isolated identities and one must introduce extra relations corresponding to these. For groupoids all coverings are co-coverings, so this last complication disappears and from a covering $\alpha : \tilde{A} \to A = \text{gpd}(X\ ; R)$ we get a presentation $\tilde{A} = \text{gpd}(\tilde{X}; \tilde{R})$, where $\tilde{R}$ consists of all pairs $(\tilde{r}_1, \tilde{r}_2)$ of edges of $\pi(\tilde{X})$ of the form $\tilde{r}_1 = (r_1, \tilde{a}),\ \tilde{r}_2 = (r_2, \tilde{a})$ with $(r_1, r_2) \in R$ and $\tilde{a}\alpha = r_1\delta = r_2\delta$.

Example 3. Normal presentations of groupoids. If in a presentation $A = \text{Gpd}(X; R)$ a relation $r = (r_1, r_2) \in R$ is such that r_1, r_2 have the same source, then we can replace r by the relation (s, e) where $s = r_1^{-1}r_2$ and e is the identity at the target of, say, r_1. A similar replacement can be made if r_1 and r_2 have the same target. If all relations are of this type then we obtain a *normal presentation* of A which we write as $A = \text{gpd}(X; S = 1)$. Here S denotes a subgraph of $\pi(X)$, and for each $s \in S$ with source i and target j, we are imposing one of the relations (s, e_i) or (s, e_j) (or both, it makes no difference). It is clear that in this case the map $\delta : \pi(X) \to A$ is a quotient map whose kernel is the normal subgroupoid of $\pi(X)$ generated by S. Applying the result of Example 2 to this case we obtain:

PROPOSITION 41. *Let* $A = \text{Gpd}(X; S = 1)$ *and let* $\alpha : \tilde{A} \to A$ *be a covering morphism in* $\mathcal{G}$. *Then* $\tilde{A} = \text{Gpd}(\tilde{X};\ \tilde{S} = 1)$, *where* $\tilde{X}$ *and* $\tilde{S}$ *are the induced covers of* X *and* S. ∎

If we now repeat the arguments of Ch.14 using this proposition we obtain a form of Reidemeister-Schreier theorem, which provides a presentation of a subgroup H of a group G when a presentation of G is known. One can also use Theorem 6, Corollary 2 (p.94). For convenience, when X is a graph and $S \subset \pi(X)$, we write $\text{gp}(X; S = 1)$ to mean the universal group of $\text{gpd}(X; S = 1)$. (The *universal group* of a groupoid A is $U_\sigma(A)$, where σ maps $V(A)$ to a one-element set). In other words, the relations imposed in $\mathrm{gp}(X;\ S = 1)$ say not only that all $s \in S$ map to identities but that all identities map to the same identity. With this notation we have:

THEOREM 14. *Let $G = \mathrm{Gp}(X;\ S = 1)$, where X is a graph with one vertex, and let H be a subgroup of G. Let $\gamma : \tilde{G} = \mathrm{Tr}(G{:}H) \to G$ be the standard covering, and let $\tilde{X}$, $\tilde{S}$ be the induced covers of X, S. Then $H = \mathrm{gp}(\tilde{X};\ \tilde{S} \cup T^* = 1)$, where T^* is obtained from a maximal tree T in $\tilde{G}$ by writing its edges as words in $\tilde{X}$. The canonical map $\tilde{X} \to H$ is constructed as follows: for each vertex i of $\tilde{G}$ there is a unique element t_i of G such that the translation $H \to Ht_i$ induced by it has target i and lies in the groupoid generated by T; if $\tilde{x} \in \tilde{X}_{ij}$ and $\tilde{x}\gamma = x$ then $\tilde{x} \mapsto t_i\, x\, t_j^{-1} \in H$.*

Proof. An exercise. ■

The same type of argument can be used to prove subgroup theorems for right limits of groups, but one cannot expect such precise information as in the case of free products. We will indicate the procedures which may give useful information in special cases. First suppose that we have a diagram in $\mathcal{G}$ with objects A^λ (and unspecified morphisms between them). If $A = \varinjlim A^\lambda$ is connected then any vertex group H of A is isomorphic with A/N, where N is a simplicial groupoid spanning A. Let A^λ have components $A^{\lambda\mu}$. Then $A = \varinjlim A^{\lambda\mu}$, where the morphisms are now those induced on components by the original ones. If $\alpha^{\lambda\mu} : A^{\lambda\mu} \to A$ are the canonical maps then $X = \bigcup A^{\lambda\mu}\alpha^{\lambda\mu}$ spans A, so we can choose a tree contained in X and spanning A. Now take N to be the groupoid generated by this tree and write $N^{\lambda\mu} = N(\alpha^{\lambda\mu})^{-1}$. The $N^{\lambda\mu}$ are normal subgroupoids of the $A^{\lambda\mu}$ and their images in A generate N as normal subgroupoid. Also the morphisms of the diagram $\{A^{\lambda\mu}\}$ induce morphisms between the $N^{\lambda\mu}$. Hence, by yet another application of Proposition 18, we have $H \cong A/N = \varinjlim A^{\lambda\mu}/N^{\lambda\mu}$.

Suppose now that $G = \varinjlim G^{\lambda}$ is a right limit of groups, and let H be any subgroup of G. Taking $A = \mathrm{Tr}(G:H)$ we have $A = \varinjlim A^{\lambda}$, where A^{λ} is the induced cover of G^{λ}. Hence we obtain as above a description of H (which is effectively a vertex group of A) as a right limit $H = \varinjlim (A^{\lambda\mu}/N^{\lambda\mu})$ of groupoids. Writing $U(C)$ for the universal group of a groupoid C, we obtain H as a right limit of groups: $H = \varinjlim U(A^{\lambda\mu}/N^{\lambda\mu})$. Now if $\overline{G}^{\lambda}$ denotes the image of G^{λ} in G, and $\overline{A}^{\lambda}$ its induced cover, then the map $A^{\lambda} \to \overline{A}^{\lambda}$ is surjective and vertex bijective, so it is a quotient map. Passing to components we have quotient maps $A^{\lambda\mu} \to \overline{A}^{\lambda\mu} \subset A$, and $A^{\lambda\mu}/N^{\lambda\mu} \cong \overline{A}^{\lambda\mu}/\overline{N}^{\lambda\mu}$, where $\overline{N}^{\lambda\mu} = N \cap \overline{A}^{\lambda\mu}$ is unicursal. Thus $A^{\lambda\mu}/N^{\lambda\mu}$ is isomorphic with a retract of $\overline{A}^{\lambda\mu}$, so is the free product of a vertex group of $\overline{A}^{\lambda\mu}$ and a free groupoid. The vertex groups of $\overline{A}^{\lambda\mu}$ are isomorphic with groups of the form $H \cap x\,\overline{G}^{\lambda}\,x^{-1}$, so we have finally $H = \varinjlim B^{\lambda\mu}$, where each $B^{\lambda\mu} = U(A^{\lambda\mu}/N^{\lambda\mu})$ has the form $(H \cap x\overline{G}^{\lambda}\,x^{-1}) * F^{\lambda\mu}$ with $F^{\lambda\mu}$ a free group. The $B^{\lambda\mu}$ will not generally be embedded in H, but their free factors $H \cap x\overline{G}^{\lambda}\,x^{-1}$ will be embedded in H since $A \to A/N$ maps vertex groups injectively.

The case of generalised free products gives most of the above information since if $A = \varinjlim A^{\lambda}$ then $A = \overline{*}\,\overline{A}^{\lambda}$, where $\overline{A}^{\lambda}$ is generated by the image of A^{λ} in A. Suppose that the group G is a generalised free product $G = \overline{*}\,G^{\lambda}$ and that H is a subgroup of G. Taking $A = \mathrm{Tr}(G:H)$ and A^{λ} the induced cover of G^{λ}, we have $A = \overline{*}\,A^{\lambda}$ (see Example 1 above). It follows that $A = \overline{*}\,A^{\lambda\mu}$, where the $A^{\lambda\mu}$ are the components of the A^{λ}. If T is a maximal tree in $\bigcup A^{\lambda\mu}$ and N the subgroupoid generated by T, then a slight modification of the above argument gives $H = \overline{*}\,H^{\lambda\mu}$, where $H^{\lambda\mu}$ is the subgroup of H

generated by the image of $A^{\lambda\mu}$ under the quotient map $A \to H$ with kernel N. Writing $N^{\lambda\mu} = N \cap A^{\lambda\mu}$ and $B^{\lambda\mu} = U(A^{\lambda\mu}/N^{\lambda\mu})$ we see that $B^{\lambda\mu}$ is of the form $(H \cap x\, G^{\lambda}\, x^{-1}) * F^{\lambda\mu}$ with $F^{\lambda\mu}$ free, and $H^{\lambda\mu}$ is a homomorphic image of $B^{\lambda\mu}$. There is a mistake (found by B. Baumslag) in [**16**], p.19 at this point, and the result stated at the bottom of that page is false. The mistake lies in assuming that (in the present notation) the map $A^{\lambda\mu}/N^{\lambda\mu} \to H^{\lambda\mu}$ is universal, i.e. that $H^{\lambda\mu} \cong B^{\lambda\mu}$. The following example (due to B.H. Neumann) shows that this is not so. Let $G = \mathrm{gp}(x, y, z;\; x^{-1} y\, x = y^{-1},\; y^{-1} z\, y = z^{-1},\; z^{-1} x\, z = x^{-1})$. Then $G = X \mathbin{\overline{*}} Y \mathbin{\overline{*}} Z$, where X, Y, Z are generated by $\{y, z\}$, $\{z, x\}$, $\{x, y\}$ respectively. Also, G has a subgroup H of order 2, generated by the element xyz. The groups X, Y, Z are torsion-free, so H meets their conjugates trivially, and the groups $B^{\lambda\mu}$ are therefore free (possibly trivial). But H is generated by their images, so they cannot all be embedded in H. The argument does, however, show that $H = \overline{*}\, H^{\lambda\mu}$, where each $H^{\lambda\mu}$ *contains* some $H \cap x_{\lambda\mu}\, G^{\lambda}\, x_{\lambda\mu}^{-1}$, and as in the Kurosh theorem the $x_{\lambda\mu}$ (for fixed λ) are a set of representatives of the double cosets $H\, x\, G^{\lambda}$. In special cases one can say more. For example, suppose that $G = G^1 \mathbin{\overline{*}} G^2$, with $G^1 \cap G^2 = G^0$, and let H be a subgroup of G such that $H \cap x G^0 x^{-1} = 1$ for all $x \in G$. Then the induced cover A^0 of G^0 in $A = \mathrm{Tr}(G : H)$ is unicursal and is the intersection of the induced covers A^1, A^2 of G^1, G^2. We therefore have $A^i = A^0 * C^i$ $(i = 1, 2)$, where C^i is a suitable retract of A^i, and clearly $A = A^0 * C^1 * C^2$. We may now choose a maximal circuit-free graph in A^0 and extend it to a maximal circuit-free subgraph T of $A^0 \cup C^1 \cup C^2$. Then T is a tree spanning A, and $H \cong A/N$, where N is generated by T. Since we are dealing with a free product, and since $N \supset A^0$, we can apply Proposition 27, Corollary, to obtain $H = H^1 * H^2$, where $H^i = U(C^i/N \cap C^i) = U(A^i/N \cap A^i)$. The full conclusion of the Kurosh theorem now follows: H is the *free product* of groups $H^{1\mu}$, $H^{2\mu}$, where $H^{i\mu} = (H \cap x_{i\mu} G^i x_{i\mu}^{-1}) * F^{i\mu}$. (See Ordman [**26**], [**27**] for further information on subgroups of generalised free products.)

Exercises

1. A category-map $\theta : A \to B$ has the unique factor-lifting property (*u.f.l.p.*) if whenever $a \in A$ and $a\theta = b_1 b_2 \dots b_n$ in B, there are unique edges $a_1, a_2, \dots, a_n$ in A such that $a = a_1 a_2 \dots a_n$ and $a_i\theta = b_i$ $(i = 1,2,\dots,n)$. Show that such a map has trivial kernel but need not be piece-wise injective. Show that a groupoid-map has *u.f.l.p.* if and only if it is a covering.

2. Prove the following extension of Theorem 13′ : if $\zeta : \tilde{Z} \to Z$ in $\mathcal{C}$ has *u.f.l.p.* (see Exercise 1) then the pull-back functor $P_\zeta : \mathcal{C}_Z \to \mathcal{C}_{\tilde{Z}}$ preserves right limits. (Prove the analogue of Theorem 8 first).

CHAPTER 16

Homology of groups and groupoids

In this section we shall describe an approach to the homology theory of groups which has a more geometric flavour than the usual algebraic method. The basic idea is to use the simplicial groupoids Δ^n in place of geometric simplexes and to imitate the singular simplicial homology theory of topological spaces as closely as possible. The method gives a homology theory for groupoids as well as for groups, and can easily be modified to apply to categories and monoids. The homology of a connected groupoid is the same as the homology of a vertex group, and for a general groupoid it is the direct sum of the homologies of the components. Thus nothing essentially new emerges; but the fact that all groupoid-maps induce maps of homology groups gives one some useful extra freedom.

We first recall some basic definitions of homological algebra. A *complex* is a family $A = \{A_n\}_{n \in \mathbf{Z}}$ of Abelian groups together with homomorphisms $\partial_n : A_n \to A_{n-1}$ such that $\partial_n \partial_{n-1} = 0$ for all n. The *homology groups* of this complex are defined as $H_n(A) = Z_n / B_n$, where $Z_n = \operatorname{Ker} \partial_n$ is the group of n-*cycles* and $B_n = \operatorname{Im} \partial_{n+1}$ is the group of n-boundaries. A is a *positive* complex if $A_n = 0$ for $n < 0$ and is a *negative* complex if $A_n = 0$ for $n > 0$. In the latter case one often writes A^n for A_{-n}, $\delta^n : A^n \to A^{n+1}$ for ∂_{-n}, Z^n for Z_{-n}, B^n for B_{-n} and $H^n(A)$ for $H_{-n}(A)$. A *morphism of complexes* $\theta : A \to A'$ is a family of homomorphisms $\theta_n : A_n \to A'_n$ such that $\theta_n \partial'_n = \partial_n \theta_{n-1}$ for all n. (One writes $\theta \partial' = \partial \theta$ for brevity). Such a morphism maps Z_n

into Z'_n, B_n into B'_n and so induces a homomorphism $H_n(\theta) : H_n(A) \to H_n(A')$, making H_n a functor from the Category of complexes to the Category of Abelian groups. If $\theta, \phi : A \to A'$ are two morphisms of complexes then a *homotopy* from θ to ϕ is a family $h = \{h_n\}$ of homomorphisms $h_n : A_n \to A'_{n+1}$ such that $h_n \partial'_{n+1} + \partial_n h_{n-1} = \theta_n - \phi_n$ for all n (abbreviated $h\partial' + \partial h = \theta - \phi$). If such a homotopy exists then $\theta_n - \phi_n$ maps Z_n into B'_n for all n, and it follows that θ and ϕ induce the same maps in homology, i.e. $H_n(\theta) = H_n(\phi) : H_n(A) \to H_n(A')$ for all n. Note also that if $h_n : A_n \to A'_{n+1}$ are arbitrary homomorphisms then $\sigma = h\partial' + \partial h$ is a morphisms of complexes and we get a homotopy from θ to $\theta - \sigma$ for any morphism $\theta : A \to A'$.

To construct homology groups for a groupoid we start by taking a standard model for the simplicial groupoid Δ^n. Its vertices are are the integers $0, 1, 2, \ldots, n$ and its edges are all pairs (i, j), where $i, j \in \{0, 1, \ldots, n\}$. In particular Δ^0 is a trivial group, and we take Δ^{-1} to be the empty groupoid. There are obvious *face-maps* $\phi^n_k : \Delta^{n-1} \to \Delta^n$ $(0 \leqslant k \leqslant n)$, where ϕ^n_k is defined to be the unique groupoid-map which sends the vertices of Δ^{n-1} *in order* to the vertices of Δ^n other than k. (Thus $i \mapsto i$ if $i < k$, and $i \to i+1$ if $i \geqslant k$).

One has the usual relations between these maps:

$$\phi^n_k \phi^{n+1}_l = \phi^n_{l-1} \phi^{n+1}_k \quad \text{if } 0 \leqslant k < l \leqslant n+1.$$

Now let G be any groupoid and define a (singular) *n-simplex in* G to be a groupoid-map $\Delta^n \to G$. If we denote the set of such n-simplexes by $\Sigma_n(G)$, then the face-maps $\phi^n_k : \Delta^{n-1} \to \Delta^n$ induce maps $\psi^k_n : \Sigma_n(G) \to \Sigma_{n-1}(G)$ satisfying

$$\psi^l_{n+1} \psi^k_n = \psi^k_{n+1} \psi^{l-1}_n \quad \text{if } 0 \leqslant k < l \leqslant n+1.$$

The Abelian groups $C_n(G)$ and $C^n(G)$ of *n-chains* and *n-cochains* in G are defined by

$$C_n = \coprod_{\sigma\in\Sigma_n} \mathbf{Z}, \quad C^n = \prod_{\sigma\in\Sigma_n} \mathbf{Z},$$

where $\mathbf{Z}$ is the additive group of the integers and $\coprod$, $\prod$ denote coproduct (direct sum) and product (Cartesian product) in the Category of Abelian groups. Thus an n-chain can be thought of as a finite collection of n-simplexes, each with an integer multiplicity (possibly negative), and an n-cochain as a map from Σ_n to $\mathbf{Z}$. The face-maps $\psi^n_k : \Sigma_n \to \Sigma_{n-1}$ induce additive maps $\partial^k_n : C_n \to C_{n-1}$ and $\delta^n_k\ C^{n-1} \to C^n$ satisfying

(1) $\quad \partial^l_{n+1}\partial^k_n = \partial^k_{n+1}\partial^{l-1}_n$ and $\delta^n_k\delta^{n+1}_l = \delta^n_{l-1}\delta^{n+1}_k$ if $0 \leqslant k \leqslant l \leqslant n+1$.

If we now define the *boundary maps* $\partial_n = \sum_{k=0}^{n} (-1)^k \partial^k_n : C_n \to C_{n-1}$, and the *coboundary maps* $\delta^n = \sum_{k=0}^{n} (-1)^k \delta^n_k : C^{n-1} \to C^n$, then the relations (1) imply $\partial_{n+1}\partial_n = 0$, $\delta^n\delta^{n+1} = 0$ $(n \geqslant 0)$, and we obtain two complexes

$$\ldots \to C_n \xrightarrow{\partial_n} C_{n-1} \to \ldots C_1 \xrightarrow{\partial_1} C_0 \to 0$$

$$0 \to C^0 \xrightarrow{\delta^1} C^1 \to \ldots \to C^{n-1} \xrightarrow{\delta^n} C^n \to \ldots$$

called the *chain complex* and *cochain complex* of G. The homology groups $H_n(G)$ and $H^n(G)$ of these complexes $(n \geqslant 0)$ are called the *homology* and *cohomology* groups of G (with integer coefficients). The groups $Z_n = \operatorname{Ker} \partial_n$, $Z^n = \operatorname{Ker} \delta^{n+1}$ are the groups of *n-cycles* and *n-cocycles* in G; the groups $B_n = \operatorname{Im} \partial_{n+1}$, $B^n = \operatorname{Im} \delta^n$ are the groups of *n-boundaries* and *n-coboundaries*. (If one extends these complexes by including C_{-1}, C^{-1} and the maps ∂_0, δ^0, one obtains the *augmented* chain and cochain complexes which give rise to the reduced homology and cohomology groups. We shall not discuss these further. Note that for the unaugmented complexes we have $H_0(G) = C_0/B_0$, $H^0(G) = Z^0$).

Any groupoid-map $\theta : X \to Y$ induces maps $\Sigma_n(\theta) : \Sigma_n(X) \to \Sigma_n(Y)$ which are compatible with the face-maps $\psi_n^k : \Sigma_n \to \Sigma_{n-1}$. These in turn induce additive maps $C_n(\theta) : C_n(X) \to C_n(Y)$ and $C^n(\theta) : C^n(Y) \to C^n(X)$ which form morphisms of complexes. Hence θ induces maps $H_n(\theta) : H_n(X) \to H_n(Y)$ and $H^n(\theta) : H^n(Y) \to H^n(X)$. One easily verifies that Σ_n is a covariant functor from $\mathcal{G}$ to $\mathcal{S}$, C_n and H_n are covariant functors from $\mathcal{G}$ to $\mathcal{A}$, and C^n, H^n are contravariant functors from $\mathcal{G}$ to $\mathcal{A}$ ($\mathcal{A}$ the Category of Abelian groups).

We shall now show that H_n and H^n are *homotopy invariants* of groupoids, i.e. if $\theta_0, \theta_1 : X \to Y$ are homotopic groupoid-maps, then $H_n(\theta_0) = H_n(\theta_1)$ and $H^n(\theta_0) = H^n(\theta_1)$ for all $n \geqslant 0$. To see this, we recall from Ch. 13 that any homotopy (natural transformation) $\theta_0 \to \theta_1$ induces a groupoid-map $\tau : X \times \Delta^1 \to Y$ such that $\mu_0 \tau = \theta_0$, $\mu_1 \tau = \theta_1$, where $\mu_0, \mu_1 : X \to X \times \Delta^1$ are the canonical embeddings. It is therefore enough to show that μ_0 and μ_1 induce the same maps in homology and in cohomology. We do this by constructing homotopies between the corresponding maps of chain and cochain complexes.

Any n-simplex $\Delta^n \to X$ induces canonically an "n-prism" $\Delta^n \times \Delta^1 \to X \times \Delta^1$, so we have a canonical map $\epsilon : \Sigma_n(X) \to P_n(X \times \Delta^1)$, where $P_n(G)$ denotes the set of n-prisms in G (i.e. groupoid-maps $\Delta^n \times \Delta^1 \to G$). By analogy with a geometric n-prism (which is an $(n+1)$-dimensional solid) we can "sub-divide" the prism $\Delta^n \times \Delta^1$ into $(n+1)$-simplexes $\sigma_k^n : \Delta^{n+1} \to \Delta^n \times \Delta^1$ $(0 \leqslant k \leqslant n)$, where σ_k^n is the unique groupoid-map which sends the vertices $0, 1, \ldots, n+1$ of Δ^{n+1} to the vertices

(0,0), (1,0), ..., $(k,0)$, $(k,1)$, $(k+1,1)$, ..., $(n,1)$, respectively, of $\Delta^n \times \Delta^1$. (This map is unique because $\Delta^n \times \Delta^1$ is again a simplicial groupoid). If we denote by $\hat{\phi}^n_l$ the "face-map" $\Delta^{n-1} \times \Delta^1 \to \Delta^n \times \Delta^1_n$ induced by the standard face-map $\phi^n_l : \Delta^{n-1} \to \Delta^n$, then the simplexes σ_k satisfy the following relations:

$$\phi^{n+1}_l \sigma^n_k = \begin{cases} \sigma^{n-1}_{k-1} \hat{\phi}^n & \text{if } 0 \leqslant l < k \leqslant n, \\ \sigma^{n-1}_k \hat{\phi}^n_{l-1} & \text{if } 0 \leqslant k < l-1 \leqslant n, \end{cases}$$

$$\phi^{n+1}_k \sigma^n_k = \phi^{n+1}_k \sigma^n_{k-1} \quad \text{if } 1 \leqslant k \leqslant n, \tag{2}$$

$$\phi^{n+1}_0 \sigma^n_0 = \nu_0\,, \quad \phi^{n+1}_{n+1} \sigma^n_n = \nu_1\,,$$

where ν_0, ν_1 are the two "ends" of the prism, i.e. the canonical embeddings $\Delta^n \to \Delta^n \times \Delta^1$. Formally, these relations imply

$$(\sum_l (-1)^l \phi^{n+1}_l)(\sum_k (-1)^k \sigma^n_k) + (\sum_k (-1)^k \sigma^{n-1}_k)(\sum_l (-1)^l \phi^n) = \nu_0 - \nu_1. \tag{3}$$

For any groupoid G, the maps $\sigma^n_k : \Delta^{n+1} \to \Delta^n \times \Delta^1$ induce maps from $P_n(G)$ to $\Sigma_{n+1}(G)$. Taking $G = X \times \Delta^1$ and composing with the canonical map $\epsilon : \Sigma(X) \to P_n(X \times \Delta^1)$ we therefore obtain corresponding maps $\tau^k_n : \Sigma_n(X) \to \Sigma_{n+1}(X \times \Delta^1)$. These in turn induce maps $\rho^k_n : C_n(X) \to C_{n+1}(X \times \Delta^1)$. The face-maps ϕ^{n+1}_l similarly induce the standard face-maps $\partial^l_{n+1} : C_{n+1}(X \times \Delta^1) \to C_n(X \times \Delta^1)$ and the maps $\hat{\phi}^n_l$ of prisms induce $\partial^l_n : C_n(X) \to C_{n-1}(X)$. The relations (2) therefore imply corresponding relations involving the ρ's and ∂'s (with order of factors reversed) and the formal relations (3) imply genuine relations

$$\rho_n \partial_{n+1} + \partial_n \rho_{n-1} = \mu^{(0)}_n - \mu^{(1)}_n,$$

where $\rho_n = \sum_k (-1)^k \rho_n^k$ and $\mu_n^{(0)}, \mu_n^{(1)}$ are the maps induced on n-chains by the embeddings $\mu_0, \mu_1 : X \to X \times \Delta^1$. Thus $\rho = \{\rho_n\}$ is a homotopy from $\{\mu_n^{(0)}\}$ to $\{\mu_n^{(1)}\}$. A similar argument applies to the corresponding maps of cochain complexes and proves

THEOREM 15. *If X is any groupoid and $\mu_0, \mu_1 : X \to X \times \Delta^1$ are the canonical embeddings, then $H_n(\mu_0) = H_n(\mu_1)$ and $H^n(\mu_0) = H^n(\mu_1)$ for all $n \geqslant 0$.* ■

COROLLARY 1. *If the groupoid-maps $\theta_0, \theta_1 : X \to Y$ are homotopic then $H_n(\theta_0) = H_n(\theta_1)$ and $H^n(\theta_0) = H^n(\theta_1)$ for all $n \geqslant 0$.*

Proof. There is a groupoid-map $\tau : X \times \Delta^1 \to Y$ such that $\theta_0 = \mu_0\tau$, $\theta_1 = \mu_1\tau$, so $H_n(\theta_0) = H_n(\mu_0) H_n(\tau) = H_n(\mu_1) H_n(\tau) = H_n(\theta_1)$, and similarly for cohomology. ■

COROLLARY 2. *Inner automorphisms of groups induce identity maps in homology and cohomology.*

Proof. Inner automorphisms of a group G are homotopic to the identity map on G. ■

COROLLARY 3. *Equivalent groupoids have isomorphic homology groups and isomorphic cohomology groups.*

Proof. If $\theta : X \to Y$ is an equivalence then for some $\phi : Y \to X$, $\theta\phi \simeq 1_X$ and $\phi\theta \simeq 1_Y$. Thus $H_n(\theta\phi)$ and $H_n(\phi\theta)$ are identity maps and $H_n(\theta)$, $H_n(\phi)$ are inverse isomorphisms. The same applies to cohomology. ■

Definition. A groupoid G is *acyclic* if $H_n(G) = H^n(G) = 0$ for $n > 0$ and $H_0(G) \cong H^0(G) \cong \mathbf{Z}$. (This definition is more reasonable in terms of reduced homology: it says that *all* the reduced homology and cohomology groups vanish).

COROLLARY 4. *Any non-empty simplicial groupoid* Δ *is acyclic.*

Proof. By Theorem 2, since Δ is connected with trivial vertex group, it is equivalent to the trivial group Δ^0. Hence it is enough to show that Δ^0 is acyclic. Now Δ^0 has exactly one n-simplex for each n, so all the groups $C_n(\Delta^0)$ and $C^n(\Delta^0)$ are isomorphic to $\mathbf{Z}$. One checks that ∂_n and δ^n are zero if n is odd, and are isomorphisms if n is even, and the result follows. ■

COROLLARY 5. *If G is any groupoid with components* G^λ, *and if* A^λ *is any vertex group of* G^λ, *then*

(i) $H_n(G) \cong \bigoplus_\lambda H_n(G^\lambda)$, $H^n(G) \cong \prod_\lambda H^n(G^\lambda)$,

(ii) $H_n(G^\lambda) \cong H_n(A^\lambda)$, $H^n(G^\lambda) \cong H^n(A^\lambda)$,

(iii) $H_0(G) \cong \bigoplus_\lambda \mathbf{Z}$, $H^0(G) \cong \prod_\lambda \mathbf{Z}$

Proof. Since the simplicial groupoids Δ^n are connected, it is clear that $\Sigma_n(G) = \coprod_\lambda \Sigma_n(G^\lambda)$, with face-maps induces by those of the separate components. Hence $C_n(G) \cong \bigoplus_\lambda C_n(G^\lambda)$, $C^n(G) \cong \prod_\lambda C^n(G^\lambda)$, and (i) follows easily since the boundary and coboundary maps respect these decompositions. Item (ii) is a consequence of Corollary 3 above and Theorem 2. Item (iii) will follow from (i) and (ii) if we can show that $H_0(A) \cong H^0(A) \cong \mathbf{Z}$ for any group A. This is left as an exercise. ■

To illustrate the fact that the homology of groupoids has interesting features not easily expressible in terms of groups, we shall now show that *universal* morphisms of groupoids induce isomorphisms in homology and cohomology in dimension $n \geqslant 2$. This can, of course, be proved by looking at the vertex groups and using known facts about the homology of free groups and free products of groups; but there may be some point in doing things the other way round. The groupoid result is a natural one and should have a direct proof. The geometric technique we adopt is that of barycentric subdivision of simplexes, which we use to construct chain and cochain homotopies in much the same way as we used subdivision of prisms above. The method is somewhat complicated and we shall not give all the details. It would be interesting to know whether there is a simpler direct proof of the theorem.

Starting with the standard simplicial groupoid Δ^n with vertices $0, 1, \ldots, n$, we take as our model for its full barycentric subdivision the simplicial groupoid $\overline{\Delta}^n$ whose vertices are all the non-empty subsets of $\{0, 1, \ldots, n\}$ (a barycentre for each subsimplex). We are interested chiefly in certain n-simplexes in $\overline{\Delta}^n$, which may be defined by specifying their vertices as follows:

(i) define $\alpha^n : \Delta^n \to \overline{\Delta}^n$ by $i \mapsto \{i\}$ $(0 \leqslant i \leqslant n)$;

(ii) for any permutation π of $\{0, 1, \ldots, n\}$, say $j \mapsto \pi_j$, define $\beta^n_\pi : \Delta^n \to \overline{\Delta}^n$ by $i \mapsto \{\pi_0, \pi_1, \ldots, \pi_i\}$ $(0 \leqslant i \leqslant n)$.

We can think of α^n, β^n_π as n-chains, i.e. elements of $C_n(\overline{\Delta}^n)$, and we now define $\beta^n \in C_n(\overline{\Delta}^n)$ by

$$\beta^n = \sum_\pi (\text{sign } \pi)\, \beta^n_\pi .$$

(β^n may be thought of as a standard subdivision of α^n).

The face-map $\phi^n_k : \Delta^{n-1} \to \Delta^n$ sends each set of vertices of Δ^{n-1} to a set of vertices of Δ^n and so induces a groupoid-map $\overline{\phi}^n_k : \overline{\Delta}^{n-1} \to \overline{\Delta}^n$. We shall refer to $\overline{\Delta}^n$ as a subdivided n-simplex and to the $\overline{\phi}^n_k$ as its subdivided faces. The formal sum $\sum_k (-1)^k\, \overline{\phi}^n_k$ induces, for each $m \geqslant 0$, a map $\overline{\partial}^n_m : C_m(\overline{\Delta}^{n-1}) \to C_m(\overline{\Delta}^n)$, and we also have the usual boundary maps $\partial^n_m : C_m(\overline{\Delta}^n) \to C_{m-1}(\overline{\Delta}^n)$. One verifies (omitting superfluous labels) that

(4) $$\alpha^n \partial = \alpha^{n-1} \bar{\partial},\ \beta^n \partial = \beta^{n-1} \bar{\partial},\ \bar{\partial}\bar{\partial} = 0,\ \bar{\partial}\partial = \partial\bar{\partial}.$$

PROPOSITION 42. *There exist chains $\eta^n \in C_{n+1}(\bar{\Delta}^n)$ $(n \geqslant 0)$ such that $\eta^n \partial + \eta^{n-1} \bar{\partial} = \alpha^n - \beta^n$ for all $n \geqslant 0$ (where η^{-1} is to be interpreted as 0).*

Proof. We use induction on n. Since $\bar{\Delta}^0$ is a trivial group we have $\alpha^0 = \beta^0$, so we may take $\eta^0 = 0$. Let $n \geqslant 1$ and suppose that η^i is defined for $0 \leqslant i \leqslant n\text{-}1$ and that $\eta^i \partial + \eta^{i-1} \bar{\partial} = \alpha^i - \beta^i$ for $0 \leqslant i \leqslant n-1$. Applying $\bar{\partial}$ to the last of these equations we get $\eta^{n-1} \partial\bar{\partial} = (\alpha^{n-1} - \beta^{n-1})\bar{\partial}$, i.e. $\eta^{n-1} \bar{\partial}\partial = (\alpha^n - \beta^n)\partial$. Thus $\alpha^n - \beta^n - \eta^{n-1}\bar{\partial}$ is an n-cylce in $\bar{\Delta}^n$. But $\bar{\Delta}^n$ is simplicial, therefore acyclic (Corollary 4 to Theorem 15), so there is an $(n+1)$-chain η^n in $\bar{\Delta}^n$ such that $\eta^n \partial = \alpha^n - \beta^n - \eta^{n-1}\bar{\partial}$, and this completes the inductive step. ■

For any groupoid G, let $\bar{\Sigma}_n(G)$ denote the set of all subdivided n-simplexes in G (i.e. all groupoid-maps $\bar{\Delta}^n \to G$), and let $\bar{C}_n(G)$, $\bar{C}^n(G)$ denote the corresponding groups of subdivided chains and cochains. The subdivided face-maps $\bar{\phi}^n_k : \bar{\Delta}^{n-1} \to \bar{\Delta}^n$ induce maps $\bar{\Sigma}_n \to \bar{\Sigma}_{n-1}$ whose alternating sums give "subdivided" boundary and coboundary maps $\bar{\partial}_n : \bar{C}_n \to \bar{C}_{n-1}$ and $\bar{\delta}_n : \bar{C}^{n-1} \to \bar{C}^n$. We thus have two new complexes $\{\bar{C}_n\}$, $\{\bar{C}^n\}$ and the relations (4) imply that the α^n and β^n induce morphisms of complexes $\alpha_*, \beta_* : \{\bar{C}_n(G)\} \to \{C_n(G)\}$ and $\alpha^*, \beta^* : \{C^n(G)\} \to \{\bar{C}^n(G)\}$. Further, the chains $\eta^n \in C_{n+1}(\bar{\Delta}^n)$ of Proposition 42 induce maps $\bar{C}_n(G) \to C_{n+1}(G)$ and $C^{n+1}(G) \to \bar{C}^n(G)$, and the proposition implies that these maps give homotopies $\eta_* : \alpha_* \to \beta_*$, $\eta^* : \alpha^* \to \beta^*$.

We define a *subdivision for* G to be an assignment, to each n-simplex $\sigma : \Delta^n \to G$ (n 0,1,...), of a subdivided n-simplex $\bar{\sigma} : \bar{\Delta}^n \to G$ in such a way that

(5) $$\alpha^n \bar{\sigma} = \sigma \text{ and } \bar{\phi}^n_k \bar{\sigma} = \overline{\phi^n_k \sigma}$$

for all n-simplexes σ and all $k=0,1,\ldots,n$. Such a subdivision gives rise to morphisms of complexes $\lambda_* : \{C_n(G)\} \to \{\bar{C}_n(G)\}$ and $\lambda^* : \{\bar{C}^n(G)\} \to \{C^n(G)\}$ such that $\lambda_* \alpha_* = 1$ and $\alpha^* \lambda^* = 1$. If we define the corresponding *subdivision chain and cochain maps* to be $s_* = \lambda_* \beta_* : \{C_n(G)\} \to \{C_n(G)\}$ and $s^* = \beta^* \lambda^* : \{C^n(G)\} \to \{C^n(G)\}$ we see that $h_* = \lambda_* \eta_*$ is a homotopy from $\lambda_* \alpha_* = 1$ to $\lambda_* \beta_* = s_*$, and $h^* = \eta^* \lambda^*$ is a homotopy from 1 to s^*. We therefore have:

PROPOSITION 43. *Any subdivision chain or cochain map for a groupoid G is homotopic to the identity and so induces the identity map in the homology or cohomology of G.* ■

Now let $\theta : A \to G$ be a universal groupoid-map. Then each edge x of G has a canonical form $x = x_1 x_2 \ldots x_n$, where the x_i are images of edges a_i of A and the products $a_i a_{i+1}$ are not defined in A (see Theorem 4, p.73). We write $|x| = n$ for the *length* of x in this reduced form; $|x| = 0$ if and only if x is an identity of G. For a simplex σ in G we define the *diameter* $|\sigma|$ of σ to be the length of its longest edge, and for a chain c in G we define the *mesh* $|c|$ of c to be the maximum of diameters of the simplexes occurring in c.

LEMMA. *There is a subdivision chain map* $s : \{C_n(G)\} \to \{C_n(G)\}$ *such that, for every* $c \in C_n(G)$, $|cs| \leqslant \frac{1}{2}(|c| + 1)$.

Proof. We have to assign to each n-simplex $\sigma : \Delta^n \to G$ a subdivided n-simplex $\bar{\sigma} : \bar{\Delta}^n \to G$ satisfying conditions (5) above. This is easily seen to be equivalent to choosing, for each n-simplex σ $(n \geqslant 1)$, an $(n+1)$-simplex σ' with $\phi^{n+1}_{n+1} \sigma' = \sigma$ (adding a barycentre). For we may then define $\bar{\sigma}$ recursively as follows: $\bar{\sigma} = \sigma$ if $n=0$, and for $n \geqslant 1$, $\bar{\sigma}$ is the unique map $\bar{\Delta}^n \to G$ whose subdivided faces are given by $\overline{\phi_k^n}\,\overline{\sigma} = \overline{\phi_k^n \sigma}$ (already defined) and such that $\mu\overline{\sigma} = \sigma'$, where $\mu : \Delta^{n+1} \to \overline{\Delta}^n$ is given by $i \mapsto \{i\}$ $(0 < i < n)$ and $n+1 \mapsto \{0,1,2,\ldots,n\}$ (the barycentre of Δ^n in $\overline{\Delta}^n$).

Our choice of barycentres is governed by the reduced forms of the edges of G. For any edge x with reduced form $x = a_1 a_2 \ldots a_n$, $a_i \in A$ (strictly, $x = (a_1\theta)(a_2\theta)\ldots(a_n\theta)$), we define a *bisection* $x = x_l x_r$ by $x_l = a_1 a_2 \ldots a_k$, $x_r = a_{k+1} \ldots a_n$, where $k = [n/2]$, with the obvious conventions if $n = 0$ or 1. Then $|x_l|$ and $|x_r|$ are both at least $\frac{1}{2}(|x| - 1)$ and at most $\frac{1}{2}(|x| + 1)$. If τ is a 2-simplex in G with edges x, y, z, say $z = xy$, we define the *median* from the edge x to be $m = x_l^{-1} x = x_r y$.

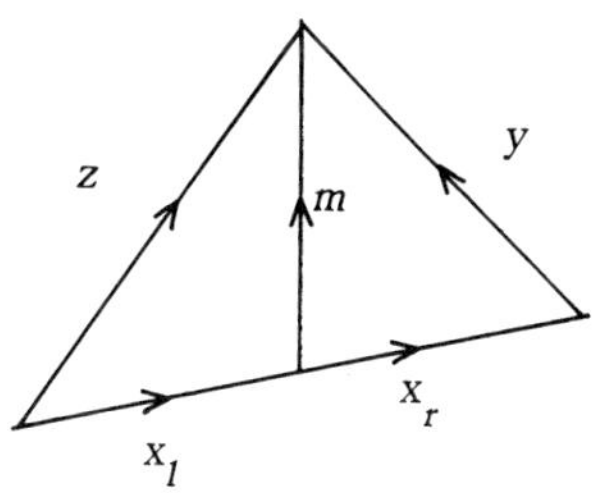

Since (in the above notation) the target of a_k is different from the source of a_{k+1}, there cannot be cancellation in both the products $x_l \cdot m$ and $x_r^{-1} \cdot m$; hence either $|z| = |x_l| + |m|$ or $|y| = |x_r| + |m|$. It follows that $|m| \leqslant max(|y|, |z|)$. Moreover, if x is a *longest* edge of τ (i.e. $|x| = |\tau|$) then $|x_l|$ and $|x_r|$ are both at least $\frac{1}{2}(|\tau| - 1)$, so $|m| \leqslant |\tau| - \frac{1}{2}(|\tau| - 1) = \frac{1}{2}(|\tau| + 1)$.

We now choose as barycentre for each simplex of G the bisector of a longest edge. More precisely, for a 1-simplex $\sigma : \Delta^1 \to G$ we regard σ as an edge of G and define σ' to be the unique 2-simplex with faces $\phi_2^2\sigma' = \sigma$, $\phi_1^2\sigma' = \sigma_l$, $\phi_0^2\sigma' = \sigma_r^{-1}$. For an n-simplex $\sigma (n \geqslant 2)$ we define σ' inductively as follows: choose an integer i such that the i th face $\phi_i^n \sigma$ of σ has maximal diameter, and let σ' be the unique (n+1)-simplex with $\phi_{n+1}^{n+1}\sigma' = \sigma$ and $\phi_i^{n+1}\sigma' = (\phi_i^n \sigma)'$. This construction defines a subdivision chain map s, and it remains to show that $|\sigma s| \leqslant \frac{1}{2}(|\sigma| + 1)$ for all n-simplexes σ. This is clear for $n=0$. For $n \geqslant 1$, let g be any edge of a simplex occurring in σs. If g already occurs in τs for some face τ of σ, then $|g| \leqslant \frac{1}{2}(|\sigma| + 1)$ by induction hypothesis, since $|\tau| \leqslant |\sigma|$. If not, then g has one end at the barycentre b of σ (which is a bisector of a longest edge) and is of one of two types: (i) g joins b to a vertex of σ; (ii) g joins b to a bisector of some edge x of σ. In case (i) g is the median from a longest edge of a 2-simplex in σ, so $|g| \leqslant \frac{1}{2}(|\sigma| + 1)$ by the argument above. In case (ii) g is the median from x in a 2-simplex whose other edges y, z meet at b and are of type (i). Since $|g| \leqslant max(|y|, |z|)$ the inequality $|g| \leqslant \frac{1}{2}(|\sigma| + 1)$ follows. ■

THEOREM 16. *If $\theta : A \to G$ is a universal groupoid-map then $H_n(\theta)$ and $H^n(\theta)$ are isomorphisms for $n \geqslant 2$.*

Proof. Let $C_n = C_n(G)$ and let $D_n \subset C_n$ be the subgroup generated by all n-simplexes in G whose diameters are 0 or 1. Clearly $\{D_n\}$ is a subcomplex of $\{C_n\}$ and we shall show that the inclusion map $\{D_n\} \to \{C_n\}$ is a homotopy equivalence (whence $\{C_n\}$ and $\{D_n\}$ have isomorphic homology groups in all dimensions). To do this it is enough to construct maps $t, k : C \to C$ ($t_n : C_n \to C_n$, $k_n : C_n \to C_{n+1}$) satisfying (i) $1\text{-}t = k\partial + \partial k$, (ii) t_n maps C_n into D_n and (iii) t_n induces the identity map on D_n. For then t, viewed as a map from C to D is a homotopy inverse to the inclusion map. We have at our disposal the subdivision map $s : C_n \to C_n$ given by the lemma and a homotopy $h : C_n \to C_{n+1}$ with $1\text{-}s = h\partial + \partial h$ given by Proposition 43. Note that h is constructed from the standard chains η^n of Proposition 42, and since s maps D to D, h will map D_n to D_{n+1}.

If σ is any n-simplex of diameter >1 then, by the lemma, $|\sigma s|<|\sigma|$. Hence some iterate $s^{(r)}$ of s maps σ into D_n. We choose $r = r(\sigma)$ to be the least such integer and define $\sigma k = \sigma(1+s+s^{(2)}+\ldots+s^{(r-1)})h$, with $\sigma k = 0$ if σ is in D_n already. This defines a map $k : C_n \to C_{n+1}$ by additivity, and if we put $t = 1\text{-}k\partial - \partial k$, we have $t : C_n \to C_n$ satisfying (i) and (iii). To show that t satisfies (ii), we observe that $\sigma(1-s)k = \sigma(1-s^{(r)})h \equiv \sigma h \pmod D$. Hence modulo D,

$$\begin{aligned}\sigma(1-s)t &= \sigma(1-s)-\sigma(1-s)k\partial - \sigma\partial(1-s)k \text{ (since } s\partial = \partial s)\\ &\equiv \sigma(1-s)-\sigma h\partial - \sigma\partial h\\ &= 0.\end{aligned}$$

Thus $\sigma t \equiv \sigma s t \pmod D$ and, by induction, $\sigma t \equiv \sigma s^{(r)} t \equiv 0 \pmod D$ for sufficiently large r.

A similar argument works for cohomology. Here we take D^n to be $\prod_\sigma \mathbf{Z}$, where σ runs through all n-simplexes of diameter 0 or 1. Then $D^n \cong \mathrm{Hom}\,(D_n, \mathbf{Z})$, $C^n \cong \mathrm{Hom}\,(C_n, \mathbf{Z})$, and applying the functor $\mathrm{Hom}(-, \mathbf{Z})$ to everything above, we obtain maps $t^n : D^n \to C^n$ which form a homotopy inverse to the canonical map $\{C^n\} \to \{D^n\}$.

Now $\theta : A \to G$ induces a morphism of complexes $\theta_* : \{C_n(A)\} \to \{D_n\}$. If we denote by $T_n(A)$ the group generated by all trivial n-simplexes in A (those with all edges equal to an identity of A), then $\{T_n(A)\}$ is a subcomplex of $\{C_n(A)\}$ and is mapped by θ_* into $\{T_n(G)\}$, which is generated by simplexes of diameter 0 in G. It is an easy consequence of the solution of the word problem for G, that every simplex of diameter 1 in G is the image of a unique (non-trivial) simplex in A. Hence θ induces an *isomorphism* of complexes $\{C_n(A)/T_n(A)\} \to \{D_n/T_n(G)\}$, hence induces isomorphisms between their homology groups. We leave it as an exercise to show that the homology groups of $\{C_n(A)/T_n(A)\}$ are, for $n \geqslant 2$, canonically isomorphic with the homology groups of $\{C_n(A)\}$ i.e. the homology groups of A. Similarly the homology groups of $\{D_n/T_n(A)\}$ are, for $n \geqslant 2$, canonically isomorphic with the homology groups of $\{D_n\}$, which we have already shown are canonically isomorphic with the homology groups of G. Hence θ induces isomorphisms $H_n(A) \to H_n(G)$ for $n \geqslant 2$, as claimed. In the case of cohomology one has complexes $\{T^n(A)\}$, $\{T^n(G)\}$ of cochains on trivial simplexes, and canonical surjections $\{C^n(A)\} \to \{T^n(A)\}$, $\{D^n\} \to \{T^n(G)\}$ whose kernels are complexes with homology groups $H^n(A)$, $H^n(G)$ in dimension $n \geqslant 2$. Again θ induces an isomorphism of complexes between these kernels and hence induces isomorphisms $H^n(G) \to H^n(A)$ for $n \geqslant 2$. ■

COROLLARY 1. *If G is a free groupoid then $H_n(G) = H^n(G) = 0$ for $n \geqslant 2$.*

Proof. There is a universal groupoid-map $\theta : A \to G$ with A unicursal (e.g. A can be an absolute free groupoid with all components isomorphic with Δ^1). Since the components of A are simplicial they are acyclic, so $H_n(A) = H^n(A) = 0$ for $n \geqslant 1$. The result follows now from the theorem. ■

COROLLARY 2. *If $G = * \, G^\lambda$ then, for $n \geqslant 2$, $H_n(G) = \oplus H_n(G^\lambda)$ and $H^n(G) = \prod H^n(G^\lambda)$.*

Proof. There is a universal groupoid-map $\theta : A \to G$ where $A = \coprod G^\lambda$. By Corollary 5 to Theorem 15, we have $H_n(A) = \oplus H_n(G^\lambda)$, $H^n(A) = \prod H^n(G^\lambda)$, and the result follows. ■

There are two directions in which the simplicial homology theory of groupoids can be extended. Firstly, one can apply the same methods to categories and monoids, replacing Δ^n by the free category on the graph $[n]$, i.e. the category with vertices $0,1,\ldots n$

and edges all pairs (i,j) with $i \leqslant j$. Theorem 15 goes through almost unchanged, but Theorem 16 requires more care, and we shall not pursue the topic further. Secondly, one can consider homology and cohomology with arbitrary coefficients (instead of the integer coefficients used so far). It is perhaps worth looking at this second generalisation briefly in the case of groupoids.

In defining the homology of a group G with general coefficients one is provided with a G-module M. i.e. a representation of G by automorphisms of an Abelian group. The corresponding notion for a groupoid G is an *additive representation* of G, i.e. a functor from G to $\mathfrak{A}$. Such a representation is given by a family $M = \{M^i\}$ of Abelian groups ($i \in V(G)$) and isomorphisms $\theta^g : M^i \to M^j$ (for edges $g \in G_{ij}$). If we forget the group structure on the M^i we get a representation $G \to \mathfrak{S}$ to which corresponds a covering $\gamma : \tilde{G} \to G$ (see Ch. 13, Proposition 30 and the preceding discussion). It is natural to consider the groups $H_n(\tilde{G})$ and $H^n(\tilde{G})$, and this would be the correct procedure if we were dealing simply with a permutational representation. But the additive structure of M gives extra structure to $\tilde{G}$ which we must take into account. Without loss of generality we can take the fibres of $\gamma : \tilde{G} \to G$ over the vertices of G to be $i\gamma^{-1} = M^i$. These are Abelian groups, and so are the fibres $g\gamma^{-1}$ over the edges of G since, for $g \in G_{ij}$, the edges covering g are in one-one correspondence (canonically) with pairs $(m,n) \in M^i \times M^j$ such that $m\theta^g = n$. Since θ^g is an isomorphism these pairs form a subgroup M^g of $M^i \times M^j$ isomorphic with M^i and M^j. We may identify the fibre $g\gamma^{-1}$ with M^g. More generally, if σ is any n-simplex in G then $\sigma\gamma^{-1}$, the set of n-simplexes in $\tilde{G}$ which cover σ, can be identified with a subgroup M^σ of $M^{i_0} \times M^{i_1} \times \ldots \times M^{i_n}$, where $i_0, i_1, \ldots, i_n$ are the vertices of σ. (Apply the corollary to Proposition 35 to maps $\Delta^n \to G$. M^σ is isomorphic with M^j for each vertex j of σ). Thus $\tilde{\Sigma}_n = \Sigma_n(\tilde{G})$ has an additive structure; it is the disjoint union of Abelian groups M^σ for $\sigma \in \Sigma_n = \Sigma_n(G)$. Furthermore, the face-maps $\tilde{\Sigma}_n \to \tilde{\Sigma}_{n-1}$ preserve this partial addition. We may therefore define modified chain and cochain groups by

$$C_n(G,M) = \bigoplus_{\sigma \in \Sigma_n} M^\sigma, \qquad C^n(G,M) = \prod_{\sigma \in \Sigma_n} M^\sigma.$$

The alternating sums of the face-maps induce boundary and coboundary maps which are group homomorphisms, and the homology groups of the resulting complexes are denoted by $H_n(G,M)$ and $H^n(G,M)$. An equivalent construction is to take the ordinary chain and cochain complexes of G and pass to the quotient complex (subcomplex in the case of cochain) obtained by imposing all the additive relations in $\tilde{\Sigma}_n$. Note that if we take the trivial representation of G on $\mathbf{Z}$ (i.e. all $M^i = \mathbf{Z}$ and all θ^g = identity) we just obtain the original groups $H_n(G)$ and $H^n(G)$.

For the case of permutational representations the situation is as follows. From a representation $G \to \mathcal{S}$ in which G acts on sets X^i we can form two natural additive representations given by families of groups $M = \{M^i\}$ and $N = \{N^i\}$, where $M^i = \bigoplus_{x \in X^i} \mathbf{Z}$, $N^i = \prod_{x \in X^i} \mathbf{Z}$. Thus M^i is the free Abelian group with basis X^i, and G acts by moving the basis elements; N^i is the set of all functions $\phi : X^i \to \mathbf{Z}$, and $g \in G_{ij}$ acts on ϕ by the rule $\phi g = \psi : X^j \to \mathbf{Z}$, where $x\psi = (xg^{-1})\phi$. If $\tilde{G}$ is the covering groupoid of G corresponding to the original representation $G \to \mathcal{S}$ then $H_n(G,M) \cong H_n(\tilde{G})$ and $H^n(G,N) = H^n(G)$. For example, if G is a group and we take the right regular representation of G, then $X = G$, and $\tilde{G}$ is the universal covering groupoid of G, which is simplicial. Since simplicial groupoids are acyclic we have $H_n(G,M) = H^n(G,N) = 0$ for $n \geqslant 1$, where now M is the group ring of G and N is the module of integer-valued functions on G.

Now consider any additive representation of a groupoid G on a family $M = \{M^i\}$ of Abelian groups, with corresponding covering $\gamma : \tilde{G} \to G$. If $\theta : K \to G$ is any groupoid-map we get an induced representation of K on the family $M_* = \{M^j_*\}$, where $M^j_* = M^{j\theta}$, $j \in V(K)$. The corresponding covering $\kappa : \tilde{K} \to K$ is the pull-back of γ by θ (Proposition 30), and we have an induced groupoid-map $\tilde{\theta} : \tilde{K} \to \tilde{G}$. It is clear that $\tilde{\theta}$ preserves all the additive structure of $\Sigma_n(\tilde{K})$, and therefore induces morphisms of complexes $\{C_n(K, M_*)\} \to \{C_n(G, M)\}$ and $\{C^n(G, M)\} \to \{C^n(K, M_*)\}$. Hence we obtain morphisms $H_n(K, M_*) \to H_n(G, M)$ and $H^n(G, M) \to H^n(K, M_*)$ induced by θ. If $\theta : K \to G$ is a *universal* morphism, then so is $\tilde{\theta} : \tilde{K} \to \tilde{G}$ (Theorem 8). We may therefore apply the barycentric subdivision argument to $\tilde{\theta}$ with a little extra care. The definition of *subdivision for* $\tilde{G}$ must be modified to include the condition that the assignment $\sigma \mapsto \bar{\sigma}$ should be compatible with the extra additive structure on $\Sigma_n(\tilde{G})$ and $\bar{\Sigma}_n(\tilde{G})$ arising from the representation. One can obtain such a subdivision for $\tilde{G}$ satisfying the required mesh inequality by lifting the given subdivision for G in an obvious way. Since $\tilde{\theta}$ preserves the addition in $\Sigma_n(\tilde{K})$, all the other constructions used in proving Theorem 16 yield maps preserving addition and therefore lead to chain maps and chain homotopies as before. It follows that θ induces isomorphisms $H_n(K, M_*) \to H_n(G, M)$ and $H^n(G, M) \to H^n(K, M_*)$ for $n \geqslant 2$. This implies that free groupoids have trivial homology and cohomology in dimension $\geqslant 2$ (for arbitrary coefficients) and that, for a free product of groupoids $G = *G^\lambda$, $H_n(G, M) \cong \oplus H_n(G^\lambda, M^\lambda)$, $H^n(G, M) \cong \prod H^n(G^\lambda, M^\lambda)$, where M^λ is the "induced module" for G^λ.

Exercises

1. If G is a group, verify that the complexes $\{C_n(G)\}$ and $\{C^n(G)\}$ defined above are isomorphic with the standard (inhomogeneous) chain and cochain complexes of G.

2. Show that if G is a group and M is a G-module, then the groups $H_n(G,M)$, $H^n(G,M)$ defined simplicially are the usual homology and cohomology groups of G with coefficients in M.

3. Use the simplicial definitions to show that if G is a group and M is any G-module then $H_n(G,M') = 0$ and $H^n(G,M'') = 0$ for $n \geqslant 1$, where $M' = M \otimes_G \mathbf{Z}G$ and $M'' = \mathrm{Hom}_G(\mathbf{Z}G,M)$. ($\mathbf{Z}G$ is the group ring of G).

4. Show that if $\theta : A \to G$ is a universal morphism of groupoids then $H_1(\theta)$ is an injection and $H^1(\theta)$ is a surjection. Show, further, that if $V(\theta)$ restricted to each component of A is an injection then both $H_1(\theta)$ and $H^1(\theta)$ are isomorphisms.

CHAPTER 17

Calculation of fundamental groups

As our last application of the theory of groupoids we shall prove a van Kampen-type theorem for fundamental groupoids which is a convenient tool for the calculation of the fundamental groups of cell-complexes. It provides another illustration of the usefulness of transferring attention from groups to groupoids; the corresponding theorem for groups is less natural and harder to apply.

We have defined in Ch. 6, for a topological space T and a subspace I, the category $P(T,I)$ of paths in T with end-points in I, and the groupoid $\pi(T,I)$ of homotopy classes of such paths. The fundamental groupoid of T is $\pi(T) = \pi(T,T)$, and the fundamental group of T at a point i is the vertex group of $\pi(T)$ at i, so knowledge of the fundamental groupoid is enough to determine the fundamental groups. Our object is to compute $\pi(T)$ from the fundamental groupoids of suitable subspaces, and we shall confine attention to the simplest case in which T is given as the union of two subspaces.

THEOREM 17. *Let T_1, T_2 be subspaces of T such that T is the union of the interiors of T_1, T_2, and let $T_0 = T_1 \cap T_2$. Then the diagram*

$$\begin{array}{ccc} \pi(T_0) & \longrightarrow & \pi(T_1) \\ \downarrow & & \downarrow \\ \pi(T_2) & \longrightarrow & \pi(T) \end{array}$$

is a pushout square in $\mathcal{G}$, the maps being those induced by the inclusion maps of spaces.

Proof. We first prove the analogous assertion for the categories of paths. Write $P(T)$ for $P(T, T)$, etc. and consider the diagram

$$\begin{array}{ccc} P(T_0) & \xrightarrow{\beta_1} & P(T_1) \\ {\scriptstyle \beta_2}\downarrow & & \downarrow{\scriptstyle \alpha_1} \\ P(T_2) & \xrightarrow[\alpha_2]{} & P(T), \end{array}$$

where the α_ν, β_ν are induced by inclusions of spaces. Let C be any category, and let $\gamma_\nu : P(T_\nu) \to C$ be category-maps such that $\beta_1\gamma_1 = \beta_2\gamma_2$. We want to construct $\gamma : P(T) \to C$ such that $\gamma_\nu = \alpha_\nu\gamma$ $(\nu = 1,2)$.

Since α_ν, β_ν are obviously injections we shall make identifications so that $P(T_0) \subset P(T_\nu) \subset P(T)$ $(\nu = 1,2)$. If $p : [0, r] \to T$ is any path in T then, since the interiors of T_1, T_2 cover T, there is (by the Lebesgue covering theorem) a positive δ such that all (closed) intervals of length $< \delta$ in $[0, r]$ are mapped either into T_1 or into T_2 by p. Hence there is a finite dissection $0 = r_0 < r_1 < \ldots < r_n = r$ of $[0, r]$ such that each $[r_{k-1}, r_k]p$ is contained in T_1 or in T_2. This gives a factorisation $p = p_1 \cdot p_2 \cdot \ldots \cdot p_n$ in $P(T)$ with each $p_k : [r_{k-1}, r_k] \to T$ lying either in $P(T_1)$ or in $P(T_2)$, and shows that $P(T)$ is generated as a category by $P(T_1)$ and $P(T_2)$. Since γ_1 and γ_2 agree on $P(T_1) \cap P(T_2)$ we may unambiguously define $p_k\gamma = p_k\gamma_\nu \in C$ if $p_k \in P(T_\nu)$. Now define $p\gamma = (p_1\gamma)(p_2\gamma)\ldots(p_n\gamma)$, observing that this product exists in C because the common end-point of p_{k-1} and p_k lies in $T_1 \cap T_2$, so is mapped to the same vertex of C by both γ_1 and γ_2. We must, of course, show that $p\gamma$ is independent of the particular dissection of $[0,r]$ chosen. If $0 = s_0 < s_1 < \ldots < s_m = r$ is another dissection with each $[s_{j-1}, s_j]p$ in T_1 or T_2, then the common refinement with dissection points all the distinct r_k, s_j has the same property, so we may assume that $\{s_j\}$ is a refinement of $\{r_k\}$. The resulting factorisation $p = q_1 \cdot q_2 \cdot \ldots \cdot q_m$ then has the property that each p_k is a product of q's lying in the same $P(T_\nu)$ as p_k. The equation $(p_1\gamma)(p_2\gamma)\ldots(p_n\gamma) = (q_1\gamma)(q_2\gamma)\ldots(q_m\gamma)$ follows since each γ_ν is a category-map.

The map $\gamma : P(T) \to C$, which is now well-defined, is obviously a category-map. Its restriction to $P(T_\nu)$ is γ_ν, and it is unique with this property since $P(T)$ is generated by $P(T_1)$ and $P(T_2)$. Hence (1) is a pushout square in $\mathcal{C}$.

Next, suppose that the category-maps $\gamma_\nu : P(T_\nu) \to C$ $(\nu = 1,2)$ are such that equivalent paths in T_ν have the same image under γ_ν (see p.32 for the definition of equivalent paths). We shall show that the induced map $\gamma : P(T) \to C$ then has the same property with respect to equivalence of paths in T. Observe that any constant path in T is a constant path in T_ν for $\nu = 1$ or $\nu = 2$ and is therefore equivalent, in T_ν, to a path of length zero, i.e. to an identity of $P(T_\nu)$. Hence, by our assumption on γ_ν, its image under γ_ν is an identity of C. Thus γ maps all constant paths in T to identities of C. If p, q are equivalent paths in T then there exist constant paths c, d such that $p' = p \cdot c$ and $q' = q \cdot d$ are of the same length and are homotopic with fixed end-points. This means that there is a continuous map $h : R \to T$, where R is the rectangle $[0, r] \times [0,1]$, which on two opposite edges of R induces the paths p', q', and on the other two edges induces constant paths a, b, as indicated in the diagram below. Again, by the Lebesgue covering theorem, there is a dissection of R into small rectangles $R_{ij} = [r_{i-1}, r_i] \times [s_{j-1}, s_j]$, where $0 = r_0 < r_1 < \ldots < r_m = r$ and $0 = s_0 < s_1 < \ldots < s_n = 1$, such that each $R_{ij}h$ is contained in T_1 or in T_2. The edges of the R_{ij} determine paths p_{ij}, a_{ij} as shown.

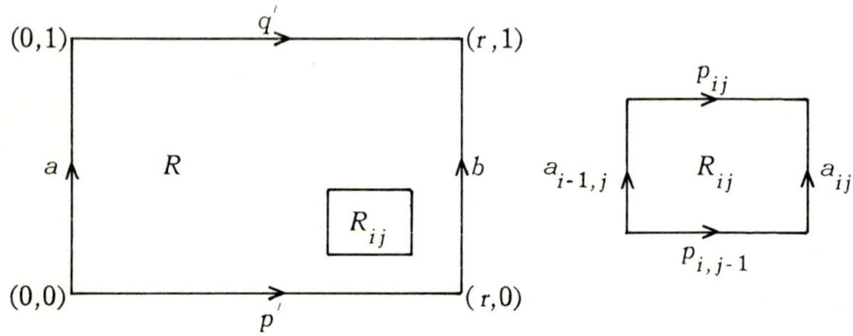

Clearly, $p' = p_{10} \cdot p_{20} \cdot \ldots \cdot p_{m0}$, $q' = q_{1n} \cdot q_{2n} \cdot \ldots \cdot q_{mn}$, $a = a_{01} \cdot a_{02} \cdot \ldots \cdot a_{0n}$ and $b = a_{m1} \cdot a_{m2} \cdot \ldots \cdot a_{mn}$. Now the rectangle R_{ij} is mapped by h into T_ν, say, and it is easy to construct from this map $R_{ij} \to T_\nu$ a homotopy in T_ν from $p_{i,j-1} \cdot a_{ij}$ to $a_{i-1,j} \cdot p_{ij}$. It follows that γ_ν (and therefore γ) maps these two paths to the same edge of C. Now consider the diagram in C consisting of all the images under γ of the paths a_{ij}, p_{ij}. It is a rectangular grid in which each small rectangle commutes, as we have just shown. It follows easily that the whole diagram commutes, and in particular, that the large rectangle consisting of the images of the edges of R commutes. Therefore $(p'\gamma)(b\gamma) = (a\gamma)(q'\gamma)$, that is $(p\gamma)(c\gamma)(b\gamma) = (a\gamma)(q\gamma)(d\gamma)$. But a, b, c, d, being constant paths, are mapped by γ to identities, and we deduce that $p\gamma = q\gamma$, as claimed.

Finally, consider the diagram

$$
(2) \qquad \begin{array}{ccc}
\pi(T_0) & \xrightarrow{\ \bar{\beta}_1\ } & \pi(T_1) \\
{\scriptstyle \bar{\beta}_2}\downarrow & & \downarrow{\scriptstyle \bar{\alpha}_1} \\
\pi(T_2) & \xrightarrow[\ \bar{\alpha}_2\ ]{} & \pi(T)
\end{array}
$$

in which the $\bar{\alpha}_\nu$, $\bar{\beta}_\nu$ are induced by α_ν, β_ν. (Note that $\bar{\alpha}_\nu$, $\bar{\beta}_\nu$ are no longer injections in general). The groupoids $\pi(T)$, $\pi(T_\nu)$ are quotients of $P(T)$, $P(T_\nu)$ by the appropriate relations of equivalence for paths, and we denote the canonical surjections by $\sigma : P(T) \to \pi(T)$ and $\sigma_\nu : P(T_\nu) \to \pi(T_\nu)$ ($\nu = 0,1,2$). If G is a groupoid and $\bar{\gamma}_\nu : \pi(T_\nu) \to G (\nu = 1,2)$ are groupoid-maps satisfying $\bar{\beta}_1 \bar{\gamma}_1 = \bar{\beta}_2 \bar{\gamma}_2$ then, for $\nu = 1,2$, $\gamma_\nu = \sigma_\nu \bar{\gamma} : P(T_\nu) \to G$ is a category-map which maps equivalent paths in T_ν to the same edge of G. Since $\beta_1 \gamma_1 = \sigma_0 \bar{\beta}_1 \bar{\gamma}_1 = \sigma_0 \bar{\beta}_2 \bar{\gamma}_2 = \beta_2 \gamma_2$, γ_1 and γ_2 induce a category-map $\gamma : P(T) \to G$ which maps equivalent paths in T to the same edge of G. Therefore there is a groupoid-map $\bar{\gamma} : \pi(T) \to G$ such that $\gamma = \sigma\bar{\gamma}$. It is easy to check that $\bar{\alpha}_\nu \bar{\gamma} = \bar{\gamma}_\nu$ ($\nu = 1,2$) and that $\bar{\gamma}$ is unique with this property, and this shows that (2) is a pushout square in $\mathcal{G}$. ■

The vertex set of $\pi(T)$ is inconveniently large for applications, so we prove a modified version of Theorem 17 with more manageable vertex sets.

THEOREM 17′. *Let T, T_0, T_1, T_2 be as in Theorem 17, and let J be a subset of T which meets every path-component of T_0, T_1 and T_0 (and therefore meets every path-component of T). Then the diagram*

$$
(3) \qquad \begin{array}{ccc}
\pi(T_0, J_0) & \longrightarrow & \pi(T_1, J_1) \\
\big\downarrow & & \big\downarrow \\
\pi(T_2, J_2) & \longrightarrow & \pi(T, J)
\end{array}
$$

is a pushout square in $\mathcal{G}$, *where* J_ν *denotes* $J \cap T_\nu$ $(\nu = 0,1,2)$.

Proof. The deduction of this result from Theorem 17 is purely algebraic. Suppose that we have a pushout square

$$
(4) \qquad \begin{array}{ccc}
G_0 & \xrightarrow{\ \theta_1\ } & G_1 \\
{\scriptstyle \theta_2}\big\downarrow & & \big\downarrow{\scriptstyle \phi_1} \\
G_2 & \xrightarrow[\ \phi_2\ ]{} & G
\end{array}
$$

in $\mathcal{G}$ with all maps vertex-injective. The vertex sets I_ν of G_ν $(\nu = 0, 1, 2)$ can be taken as subsets of $I = V(G)$, and then $I_1 \cup I_2 = I$, $I_1 \cup I_2 = I_0$. Let J be a subset of I containing at least one vertex of each component of G_0, G_1 and G_2, and let H, H_ν be the full subgroupoids of G, G_ν with vertex sets J, J_ν, where $J_\nu = J \cap I_\nu$. Then (4) induces a commutative diagram

$$
(5) \qquad \begin{array}{ccc}
H_0 & \xrightarrow{\ \theta_1'\ } & H_1 \\
{\scriptstyle \theta_2'}\big\downarrow & & \big\downarrow{\scriptstyle \phi_1'} \\
H_2 & \xrightarrow[\ \phi_2'\ ]{} & H
\end{array}
$$

which, as we shall show, is also a pushout square in $\mathcal{G}$. The theorem then follows on putting $G = \pi(T)$, $G_\nu = \pi(T_\nu)$. By Theorem 2, H, H_ν are retracts of G, G_ν, and we shall construct retractions

$\rho : G \to H$, $\rho_\nu : G_\nu \to H_\nu$ which form a morphism of diagrams from (4) to (5). First choose for each $i \in I_0$ an edge x_i^0 in G_0 from i to some vertex $j \in J_0$, with $x_i^0 = e_i$ if $i \in J_0$. These x_0^i generate a unicursal normal subgroupoid N_0 of G_0 which is the kernel of a retraction $\rho_0 : G_0 \to H_0$. Next for each $i \in I_\nu$ ($\nu = 1,2$) choose an edge x_i^ν in G_ν from i to some $j \in J_\nu$ such that (i) $x_i^\nu = e_i$ for $i \in J_\nu$ and (ii) $x_i^\nu = x_i^0 \theta_\nu$ if $i \in I_0$. Again we have unique retractions $\rho_\nu : G_\nu \to H_\nu$ with kernels N_ν generated by the x_i^ν. We may now define, for each $i \in I$, $x_i = x_i^\nu \phi_\nu$ if $i \in I_\nu$, and so obtain a retraction $\rho : G \to H$ with kernel N generated by the images of N_1, N_2. If we think of G as the right limit of the G_ν with maps θ_1, θ_2 we are in precisely the situation described by Proposition 27 (i) (quotients commute with right limits). We conclude that $H \cong G/N$ is the right limit of the $H_\nu \cong G_\nu/N_\nu$ with respect to the appropriate maps, and it is clear that the appropriate maps are those in (5). It follows that (5) is a pushout square. (Another argument is indicated in Excercise 1 below). ■

Note. One might hope that this algebraic argument would work for arbitrary right limits, but there are difficult combinatorial problems involved in the choice of retractions. Some general conditions under which a coherent choice is possible are given in [**5**]. Perhaps a study of the question would throw light on the difficulties encountered in Ch.15 when attempting to prove subgroup theorems for right limits of groups.

We end by indicating how Theorem 17′ is used to calculate fundamental groups in simple cases. If a space X is contractible (i.e. 1_X is homotopic to a constant map $X \to X$) then X is homotopy equivalent to a one-point space, so $\pi(X)$ is equivalent to a trivial group, that is, $\pi(X)$ is simplicial. This is the case, for example if

X is the unit interval $I = [0,1]$ or the n-cube I^n. If a space T can be covered by the interiors of a finite number of contractible spaces which intersect decently then repeated application of the theorem yields a presentation of $\pi(T,J)$ for suitable $J \subset T$. For if we know presentations of $\pi(T_\nu, J_\nu)$ for $\nu = 1,2$ and a set of generators for $\pi(T_0, J_0)$, the pushout square (3) immediately gives a presentation of $\pi(T, J)$. If T is path-connected one can then obtain a presentation of its fundamental group by retracting $\pi(T, J)$ to a vertex, which is equivalent to adding relations $R = 1$, where R is a maximal tree (see Theorem 6, Corollary 2, p.94).

The simplest non-trivial case is that of a circle S^1. It can be covered by the interiors of two copies T_1, T_2 of the unit interval I whose intersection T_0 has two path-components each homeomorphic to I. If we take for J a pair of points, i,j, one in each component of T_0, then the groupoids $G_1 = \pi(T_1, J)$, $G_2 = \pi(T_2, J)$ are both simplicial with two vertices (i.e. isomorphic with Δ^1) and $G_0 = \pi(T_0, J)$ is the trivial groupoid with two vertices. If we let x_1 be the edge of G_1 from i to j and x_2 the edge of G_2 from i to j then G_1, G_2 are freely generated by x_1, x_2 respectively. It follows from Theorem 17$'$ that $G = \pi(T, J)$ is freely generated by the images y_1, y_2 of x_1 and x_2. If we now retract to the vertex i by imposing the relation $y_2 = 1$ we see that the fundamental group of S^1 is a free group on one generator, i.e. an infinite cyclic group.

The cylinder $S^1 \times I$ is homotopy equivalent to S^1, so also has infinite cyclic fundamental group. This space can be used to compute the fundamental groups of the torus $S^1 \times S^1$ and the Klein bottle K, each of which can be covered by the interiors of two subspaces T_1, T_2 homeomorphic with $S^1 \times I$ so that the intersection $T_0 = T_1 \cap T_2$ has two components each homeomorphic with $S^1 \times I$. We take for J a pair of points i,j, one in each component of T_0. The groupoid $G_1 = \pi(T_1, J)$ is then a free groupoid on two generators which we may take as edges $x_1, y_1 : i \to j$. The two elements $a_1 = x_1 y_1^{-1}$ and $b_1 = y_1^{-1} x_1$ are then free generators of the two vertex groups at i and j, respectively, and they can be represented by loops round T_1 in the same direction since they are conjugate ($b_1 = y_1^{-1} a_1 y_1$). Similarly $G_2 = \pi(T_2, J)$ is freely generated by $x_2, y_2 : i \to j$, and its vertex groups at i, j are freely generated by $a_2 = x_2 y_2^{-1}$ and $b_2 = y_2^{-1} x_2$, respectively. $G_0 = \pi(T_0, J)$ is the disjoint union of two infinite cyclic groups at i, j, generated by a_0, b_0, say.

In the case of the torus $S^1 \times S^1$ the end segments of T_1 and T_2 are overlapped so that loops in T_0 at i and j which go round T_1 in the same sense, also go round T_2 in the same sense. Thus we may define the map $G_0 \mapsto G_1$ by $a_0 \mapsto a_1$, $b_0 \mapsto b_1$ and the map $G_0 \to G_2$ by $a_0 \mapsto a_2$, $b_0 \mapsto b_2$. The pushout $G = \pi(S^1 \times S^1, J)$ then has generators x_1, y_1, x_2, y_2 with defining relations $a_1 = a_2$ and $b_1 = b_2$, i.e. $x_1 y_1^{-1} = x_2 y_2^{-1}$ and $y_1^{-1} x_1 = y_2^{-1} x_2$. If we retract to the vertex i by means of the tree consisting of y_2, say, we obtain a presentation of the fundamental group of $S^1 \times S^1$ in the form $\text{gp}\{x_1, y_1, x_2 ; x_1 y_1^{-1} = x_2, y_1^{-1} x_1 = x_2\}$ (just put $y_2 = 1$ in the given presentation of G). This is equivalent to $\text{gp}\{x_1, y_1 ; x_1 y_1 = y_1 x_1\}$, which is the direct product of two infinite cyclic groups. This conclusion could also be reached by Exercise 4 of 4, p.59. The actual generators of the vertex group at i are the images of x_1, y_1 under the retraction, namely $\zeta = x_1 y_2^{-1}$ and $\eta = y_1 y_2^{-1}$.

In the case of the Klein bottle K, the ends of T_1 and T_2 overlap in opposite senses, so we may define $G_0 \to G_1$ by $a_0 \mapsto a_1$, $b_0 \mapsto b_1$ and $G_0 \to G_2$ by $a_0 \mapsto a_2$, $b_0 \mapsto b_2^{-1}$. The pushout $\pi(K, J)$ now has generators x_1, y_1, x_2, y_2 with defining relations $x_1 y_1^{-1} = x_2 y_2^{-1}$ and $y_1^{-1} x_1 = (y_2^{-1} x_2)^{-1}$. Retracting to i by putting

$y_2 = 1$ we get a presentation of the fundamental group of K in the form $\mathrm{gp}\{x_1, y_1, x_2 ; x_1 y_1^{-1} = x_2, y_1^{-1} x_1 = x_2^{-1}\}$, which is equivalent to $\mathrm{gp}\{x_1, y_1 ; x_1^2 = y_1^2\}$. Again the corresponding generators of the fundamental group at i are $\xi = x_1 y_2^{-1}$, $\eta = y_1 y_2^{-1}$.

Exercises

1. In any Category $\mathcal{K}$, let $\alpha : \mathbf{A} \to \Gamma(A)$, $\beta : \mathbf{B} \to \Gamma(B)$ be morphisms of D-diagrams, where $\Gamma(A)$ denotes the constant diagram at the object A of $\mathcal{K}$. Suppose that θ is a morphism from α to β (i.e. a pair of morphisms $\theta_1 : \mathbf{A} \to \mathbf{B}$ and $\theta_2 : \Gamma(A) \to \Gamma(B)$ such that $\theta_1 \beta = \alpha \theta_2$) and that θ has a left inverse ϕ (a morphism $\phi : \beta \to \alpha$ such that $\phi\theta$ is the identity morphism on β). Prove that if α is a right limit for $\mathbf{A}$ then β is a right limit for $\mathbf{B}$. Use this result to complete the proof of Theorem 17$'$ instead of using Proposition 27.

2. Prove that the n-sphere S^n has trivial fundamental group for $n \geqslant 2$.

3. Prove that the fundamental group of S^2 with k points removed is a free group on $k-1$ generators.

4. Find the fundamental group of the complement T in real 3-space R^3 of the knotted curve C below.

(Hint. Take two neighbouring parallel planes H_1, H_2 perpendicular to the plane of the page each cutting C in four points. Let U_1 U_2

be open half-spaces bounded by H_1, H_2 with $U_1 \cap U_2$ the region between H_1 and H_2. Take $T_i = U_i \cap T$, a half-space with two arcs removed.)

5. Let X be a graph with vertex set V, edge set E and incidences $\delta_0, \delta_1 : E \to V$. Let $(0,1)$, $[0,1]$ denote open and closed real unit intervals. Let $|X| = (E \times (0,1)) \cup V$ (disjoint union), and topologise X as follows. Give E the discrete topology, $E \times [0,1]$ the product topology, and $|X|$ the identification topology induced by the obvious map $E \times [0,1] \to |X|$ $((x,0) \to x\delta_0 \in V,\ (x,1) \to x\delta_1 \in V,\ (x,\lambda) \to (x,\lambda)$ for $0 < \lambda < 1)$. Prove that $\pi(|X|, V) \cong \pi(X)$ (natural equivalence of functors).

Bibliographical Notes

All the category theory contained in these pages is to be found in the textbooks of Freyd [13], Mitchell [22] and Bucur and Deleanu [8]. The reader who wants to pursue the subject further should consult these and the excellent survey by MacLane [20].

It is interesting to note that groupoids appeared in the literature nearly twenty years before categories. They were introduced by Brandt [3], who discovered them in his study of composition of quadratic forms and used them again in [4] to describe multiplication of ideals in orders over Dedekind domains. At about the same time Loewy [19] introduced similar "compound groups" to describe isomorphisms between conjugate field extensions. His ideas were developed by Baer in [2]. Apart from a small number of passing references, the two concepts seem to have been quickly forgotten, probably because of a general distaste for partial operations. It was not until categories were generally accepted (some ten years after their introduction by Eilenberg and MacLane [11] in 1945) that interest in groupoids revived. Their systematic study was initiated by Ehresmann in a long series of papers on local structures summarised in [12]. They have also been used by Dedecker to describe cohomology with non-Abelian coefficients (see [10] and the references given there) and by Michler [21], who continued, in the language of groupoids, the work of Loewy and Baer on Galois theory.

The applications to group theory described in Chs. 14, 15 above are due to M. Hasse [15] and the author [16, 17]. The methods are derived on the one hand from the graph-theoretical methods initiated by Cayley (see Burnside [9], Ch. XIX) and generalised by Reidemeister [23], and on the other hand from the related method of covering spaces. More recently, Serre [24] has given a similar treatment of the Kurosh theorem using the notion of a group acting on a graph. For a topological proof of Grushko's theorem, see Stallings [25].

The simplicial approach to the homology of groups, described in Ch.16, was suggested by C. Rourke (unpublished), and is closely related to the work of André [1] and others. The application of groupoids to homotopy theory has been nicely exploited by R. Brown in his book [6], and it is his version of the van Kampen theorem which we have described in Ch.17. In [5], Brown proves a stronger "adjunction theorem" and also treats the interesting case of an infinite union of spaces.

Suggested further reading: André [1], Brown [5, 6, 7], Gabriel and Zisman [14], Serre [24].

References

1. ANDRE, M. *Méthode simpliciale en algèbre homologique et algèbre commutive* (Springer, 1967)

2. BAER, R. Beiträge zur Galoisschen Theorie, *S.-B. Heidelberger Akad. Wiss. Math. Nat. K1., 1928, Abh. 14.*

3. BRANDT, H. 'Uber ein verallgemeinerung des Gruppenbegriffes', *Math. Ann.* **96**, 360-366 (1926)

4. BRANDT, H. 'Idealtheorie in einer Dedekindsche Algebra', *Jber. Deutsch. Math. Verein.* **37** 5-7 (1928)

5. BROWN, R. 'Groupoids and van Kampen's theorem', *Proc. Lond. Math. Soc. (3)* **17** 385-401 (1967)

6. BROWN, R. *Elements of modern topology* (McGraw-Hill, 1968)

7. BROWN, R. *Fibrations of groupoids, J. Algebra* **15**, 103-132 (1970)

8. BUCUR, I. and DELEANU, A. *Introduction to the theory of categories and functors* (Interscience, 1968)

9. BURNSIDE, W. *Theory of groups of finite order* (Cambridge, 1911) (Reprinted by Dover, 1955)

10. DEDECKER, P. 'Sur la cohomologie non abélienne II', *Can. J. Math.* 15, 84-93 (1963)

11. EILENBERG, S. and MACLANE, S. 'The general theory of natural equivalences', *Trans. Am. Math. Soc.* 58, 231-294 (1945)

12. EHRESMANN, C. '*Catégories et structures* (Dunod, Paris, 1965)

13. FREYD, P. *Abelian categories* (Harper and Row, 1964)

14. GABRIEL, P. and ZISMAN, M. *Calculus of fractions and homotopy theory* (Springer-Verlag, 1967)

15. HASSE, M. 'Einige bemerkung über Graphen, Kategorien und Gruppoide', *Math. Nachr.* **22**, 255-270 (1960)

16. HIGGINS, P.J. 'Presentations of groupoids, with applications to groups', *Proc. Camb. Phil. Soc.* **60**, 7-20 (1964)

17. HIGGINS, P.J. 'Grushko's theorem', *J. Algebra* **4**, 365-372 (1966)

18. KUROSH, A. *The theory of groups*, Vol. II. English translation, Chelsea, 1955.

19. LOEWY, A. 'Neue elementare Begründung und Eweiterung der Galoisschem Theorie', *S.-B. Heidelberger Akad. Wiss. Math. Nat. K1. Abh.* 7 (1925); *Abh.* 1 (1927)

20. MACLANE, S. 'Categorical algebra', *Bull. Amer. Math. Soc.* **71**, 40-106 (1965)

21. MICHLER, L. 'Uber eine Verallgemeinerung des Hauptsatzes der Galoisschen Theorie', *Wiss. Z. Hochsch. Schwermaschinenbau Magdeburg*, I, *Heft* 1 (1956-57)

22. MITCHELL, B. *Theory of categories* (Academic Press, 1965)

23. REIDEMEISTER, K. *Einführung in die kombinatorische Topologie*. (Brunswick, 1932)

24. SERRE, J.-P. *Groupes discrètes*. (Lecture notes to be published)

25. STALLINGS, J.R. 'A topological proof of Grushko's theorem on free products', *Math. Z.* **90**, 1-8 (1965)

26. ORDMAN, E. T. *Amalgamated free products of groupoids*. (Dissertation, Princeton University, 1969)

27. ORDMAN, E. T. 'On subgroups of amalgamated free products', *Proc. Camb. Phil. Soc.* (to appear).

INDEX